THE ORIGINS OF AGNOSTICISM

THE ORIGINS OF AGNOSTICISM

VICTORIAN UNBELIEF AND THE LIMITS OF KNOWLEDGE

BERNARD LIGHTMAN

THE JOHNS HOPKINS UNIVERSITY PRESS
Baltimore and London

This book has been brought to publication
with the generous assistance of
the Andrew W. Mellon Foundation.

Printed in the United States of America

The Johns Hopkins University Press
701 West 40th Street
Baltimore, Maryland 21211
The Johns Hopkins Press Ltd., London

The paper used in this publication
meets the minimum requirements of American
National Standard for Information Sciences—
Permanence of Paper for Printed Library
Materials, ANSI Z39.48-1984.

Library of Congress Cataloging-in-Publication Data

Lightman, Bernard V., 1950–
The origins of agnosticism.

Bibliography: p.
Includes index.
1. Agnosticism—History—19th century.
2. Great Britain—Intellectual life—19th century.
I. Title.
BL2747.2.L54 1987 211'.7'09034 87-2690
ISBN 0-8018-3375-2 (alk. paper)

CONTENTS

ILLUSTRATIONS

ACKNOWLEDGMENTS

This study of agnosticism would not have seen the light of day without the generous support of various institutions and the encouragement of teachers and friends. Through the Contingency Fund of the University of Oregon Foundation I received help in the final preparation of the manuscript for publication. I am grateful to the School of Graduate Studies and Research at Queen's University for both a research travel grant and an Advisory Research Committee award. I also wish to thank the Social Sciences and Humanities Research Council of Canada, whose grant of a postdoctoral fellowship provided the time in which to write the book. The kindly interest of Peter Blaney of the Canadian Federation of the Humanities' Aid to Publications Programme gave me the strength to endure several rounds of revisions.

I was fortunate in finding cheerful and competent assistance from the interlibrary loan staff of both York University and Queen's University, who surely dreaded my all-too-regular visits. I must also acknowledge my gratitude to Mrs. McCabe of the Royal Institution, London, and Mrs. Jeanne Pingree of the Imperial College of Science and Technology, London, for their helpfulness when I worked on the Tyndall and Huxley papers. I appreciate receiving permission from the following institutions and individuals to quote from manuscript collections, enabling me to present a richer picture of agnosticism: Professor Quentin Bell, the Imperial College Archives (Huxley Papers), the Director of the University of London Library and the Athenaeum, Pall Mall, London (Herbert Spencer Papers), the University College London Library, and the Bodleian Library. Material from the Tyndall Papers appears by courtesy of the Royal Institution.

I owe a still deeper debt to my colleagues and teachers. It may seem extravagant to thank an old high-school history teacher, but I regarded Ernie Brown with awe and reverence in those days. I have no doubt that he was the first to awaken in me a respect for history, which was carefully cultivated by William Magney at York University. I first seriously tackled agnosticism in an essay for a graduate course on Victorian religion taught by Richard Helmstadter of the University of Toronto and Norman Feltes of York University. Bill Gay and Sam Schweber, who served on my doctoral thesis committee, will forever remind me of happy days in the history of ideas program at Brandeis.

I must also express my appreciation to my friends Brady Polka, Paul Christianson, James Moore, Ruth Barton, and Paul Fayter, who have read parts of the dissertation or the manuscript and have shared with me their stimulating comments. Words of encouragement from Frank Turner have helped me to persevere. John W. Bicknell has kindly fed me choice tidbits of information about Leslie Stephen throughout a correspondence in which I tried his patience with an infinite number of questions.

Three teachers have been special to me in the course of my education: Syd Eisen, who first introduced me to Victorian intellectual history and continues to act as adviser, teacher, and colleague; Bill Johnson, my thesis adviser at Brandeis and guide through the troubled waters of graduate life; and finally, Brady Polka, who showed me that intellectual history was inseparable from the process of thinking through one's life. My pen fails me when I try to convey my sense of love and obligation to my family, who have always supported my work even if they were not quite sure what it all really meant. My long-suffering wife, Merle, has borne the full impact of a dissertation and a book with astonishing cheerfulness. This book is dedicated to all of you: my parents, John, Ryla, Romy, Sarah, Art, E. J., Jordy, Merle, and my son, Matthew.

THE ORIGINS OF AGNOSTICISM

Introduction

THE POWER OF MODERN AGNOSTICISM

A leading English Positivist and advocate of the Religion of Humanity, Frederic Harrison (1831–1923), was in a unique position to prophesy on "The Future of Agnosticism." In addition to maintaining personal friendships with the leading agnostics of the day, he could sympathize with their attack on traditional theology and vague metaphysics. However, his commitment to Auguste Comte led him to adopt an attitude favorable to some type of organized religion, and that position placed him outside the pale of agnosticism. In the pages of the *Fortnightly Review* for January 1889, Harrison took on the role of seer and presented his vision of the future in order to determine whether the widespread agnostic phase of mind could be permanent, final, and creative. He asked, "Is Agnosticism a substantive religious belief at all? Can it grow into a religious belief? Can it supersede religious belief?" Harrison concluded that agnosticism as a "creed" would not stand the test of time, that it had no future due to its purely destructive character. Agnosticism was "a state of no-religion," and since man was by nature a religious animal, the inadequacy of the agnostic position would eventually be discovered (pp. 144, 149).

But although Harrison's crystal ball told him that agnosticism as a distinct school of thought would vanish completely, he also predicted, paradoxically, that "agnostic logic" would become universally accepted as part of our intellectual baggage. As "minds are more commonly imbued with the sense of physical law," Harrison decreed, agnostic logic was bound to become an "axiom of ordinary thought, almost a truism or a commonplace" (p. 154). Although Harrison's pessimism about the survival of agnosticism as a distinct faith obviously serves as a convenient pretext for displaying the superior charms of his lady liege Positivism, his assessment of agnosticism's fate was surprisingly accurate. Agnosticism as a "creed" died with Leslie Stephen, in 1904, while the power of agnostic assumptions lived on in the early twentieth century in various forms, sometimes appearing in humanist philosophies, at other times in the shape of positivism or secularism.

It is not often these days that we find intellectuals willing to call themselves, first and foremost, agnostics. But, in general, agnosticism today represents a pervasive yet diffuse attitude that has moved in quite a different direction when we recall the self-assurance of agnostics such as T. H. Huxley when they confidently proclaimed their position. Baumer has perceptively portrayed the strange situation in which twentieth-century agnosticism finds itself. Similar to other types of scepticism, it "has now become a problem where once it seemed a release and a relief."[1] Baumer calls our century "The Age of Longing" (after Arthur Koestler's novel) in order to symbolize how irreligious scepticism has combined in a new way with a longing for the God who is dead or for a God not yet born. Where Huxley exulted in his feeling of liberation from the oppressive bonds of Christian faith, Baumer sees in the despair of modern scepticism an agonizing sense of loss and an awareness that, in spite of its logical consistency, religious unbelief makes life seem meaningless and hollow.

A poignant case is presented by the anthropologist Bronislaw Malinowski, one of the few modern sceptics who is a self-proclaimed agnostic. In a symposium on science and religion in 1931, Malinowski refers to modern agnosticism as "a tragic and shattering frame of mind."[2] There can be no doubt that Baumer is onto something of immense importance here. Many of the convictions held by Victorian agnostics which gave their agnosticism its vitality and forward-looking optimism have not survived the shock of two great wars, the inhumanity of Fascism, and the threat of nuclear destruction. The texture and very existence of the original agnosticism was closely bound up with the spirit of the Victorian age and the continued well-being of liberalism. Huxley's almost naive belief in progress and his faith in the potential of science can scarcely be accepted today. Agnosticism has become tragic—a position forced upon us against our wills by a twentieth-century worldview steeped in scientific assumptions. "Is science responsible for my agnosticism," Malinowski asks, "and for that of others who think like me? I believe it is, and therefore I do not love science, though I have to remain its loyal servant."[3] What a terrible confession to wring from the lips of an anthropologist, and what a dilemma for modern man! Modern agnosticism is ingrained in our consciousness, according to Harrison, and seems to confront us with two equally unattractive alternatives. Either we unlovingly accept science and the agnosticism it apparently demands, thereby alienating ourselves from the spiritual world of religion, or we give up the search for knowledge and rush unthinkingly to embrace religion.

The forces that first led to the birth of agnosticism and then later transformed it almost beyond recognition have created numerous pitfalls in the way of undertaking a historical study of agnosticism. There is a special difficulty in interpreting the origins of agnosticism which requires the historian to become conscious of his own presuppositions about the world. The twentieth-century North American and British worldview, which is molded by the rise of evolutionary theory, the development of agnosticism, and the pervasiveness of empiricism, is the very obstruction blinding us to the story of the origins of agnosticism. We hold in common so much with the agnostic frame of mind that it is necessary to distance ourselves from it in order to perceive the historical significance of the agnostics' thought.

Ultimately we share just enough with the original agnostics to be confronted with two related problems. First, the element of commonality can lead us to flatten down the diverse and protean quality of Victorian agnosticism into a completed picture. The 1860s was an important decade for the formation of the modern religious consciousness. However, there was much in the thought of that decade which is foreign to us today. We look back upon the Victorian period with the benefit of the completed process of the development of agnosticism, and we forget how much the nineteenth century was an age of transition. As a result we de-emphasize the religious quality of agnosticism, the ambivalent attitude of agnostics toward Darwin's theory of evolution, and their metaphysical idealism. These elements have been extracted from modern agnosticism. As rigorous empiricists, we ignore the Victorian agnostics' idealism. Inasmuch as Darwin's theory of natural selection has become scientific orthodoxy, we must remind ourselves that Darwin was not fully vindicated until the early twentieth century and that the agnostics had reservations about accepting Darwinism as the final word on evolution. Due to our tendency to view agnosticism primarily as an antireligious mode of thought, we find it difficult to entertain either the notion that there were many vestiges of traditional religious thought embedded in Victorian agnosticism or the possibility that agnosticism originated in a religious context. Another aspect of this same difficulty is our propensity to view Victorian agnosticism as monolithic in nature. Actually, there were a number of types of agnosticism, including theistic and atheistic varieties.

A second problem concerns our inability to get outside the alleged empiricism of Victorian agnosticism in order to see through its position. In order to present my perspective on agnosticism I will be obliged to undertake a critique of the empiricism that we have inherited from late-nineteenth-century England. In his historical reconstruction of the past, Popkin has also committed himself to a rejection of "most Anglo-American philosophy," because he perceives in it an atmosphere of triviality that comes from its refusal to face up to the implications of Hume's work. "To get beyond Hume," Popkin argues, "would require some basis for guaranteeing or justifying our knowledge that showed that we could somehow know the nature of reality. The British answer, in failing to come to grips with the basic epistemological issues, has left British philosophy adrift ever since, vascillating between reporting what we have to believe, how we speak, etc., and making a virtue of Humeanism in the form of positivism."[4] We must then look outside the English empiricist tradition for a sophisticated position from which to evaluate the agnostics.

As is well known, it was Hume who awoke Kant from his dogmatic slumber. It is only appropriate that we make use of the Kantian perspective in order to transcend the empiricism of Hume and the agnostics. By the Kantian perspective I mean a position that seeks to overcome the fatal opposition between science and religion presented by the powerful but tragic twentieth-century sceptic. Ever since the time of the ancient Greeks, the Western tradition has been confronted by a series of false dualisms of which the science-religion dichotomy is only one example. Body has been opposed to soul, matter to mind, phenomena to noumena, necessity to freedom. Kant wished to approach these so-called opposites in such a way as to make a choice between one or the other unnecessary. He was not prepared to give up either side of each dualism, as he recognized that doing so repressed a significant dimension of human life. By making a critical examination of the extent of reason's powers, and by applying his subsequent conclusion that reason has the ability to ground science as well as to find religious truth, he aimed to preserve both science and religion in their full integrity. Kant has no monopoly on this approach. Other great minds have also attempted a similar project.[5] Indeed, the agnostics themselves claimed to have reconciled science and religion; however, unlike Kant, they failed to reach a consistent position. To single out Kant for special treatment in our study of agnosticism, from among all those who sought unity in human life, is not at all arbitrary and does not require the inclusion of extraneous material. For it is legitimate, as we shall see, to trace agnostic epistemology back to the philosopher of Königsberg. If we are to allow Kant's texts to engage us on all levels, both as documents that are part of the chronology of agnosticism *and* as works of literature that address us, we cannot neglect what we learn from the questions he raises for us. The re-creative dialogue with the texts of a great thinker should transform our perspective on the origins of agnosticism.[6]

The last major work on Victorian agnosticism, Cockshut's *Unbelievers*, was published in 1966. However, studies of the past decade, by Young, Gillespie, and Moore, have shown that, from the point of view of intellectual history, the Victorian age was much more of a transitional period than was previously thought. Young has pointed to the line of continuity linking the Paleyan tradition with scientific naturalism, Gillespie has made us more aware of the remnants of religious thought embedded in Darwin's mind, and Moore has forced us to re-examine the customary way of viewing the relationship between science and religion through the metaphor of warfare. The catastrophist position in Victorian intellectual history, that 1859 represents a gigantic upheaval in the English philosophical framework, may be losing the struggle for existence. The significant implications of this new literature for our understanding of Victorian unbelief signal the need for new studies of agnosticism which preserve the complexity and continuity of the process of change in Victorian thought patterns.

A beginning effort toward a reinterpretation of agnosticism must deal with the central claim made by all of the original agnostics: that God is unknowable. This is an epistemological assertion that demands of the historian an understanding of the roots of the theory of knowledge constructed by Huxley and his fellow agnostics. We must focus our attention here rather than dwell on the impact of evolutionary theory or the influence of biblical criticism. If we look carefully into the sources of the agnostics' stress on the limits of knowledge, we will find ourselves face to face with the strange discovery that agnosticism owes a profound debt to an epistemological position put forward by a number of ardent Christian thinkers. This is more than a quirk of intellectual history; it points to the religious origins of agnosticism.

Harrison was quite perceptive when he predicted that the agnostic frame of mind would become more prevalent in the twentieth century even as a distinctive agnostic faith would all but disappear. He would have been amused to find that the future he envisioned for agnosticism has erected barriers against attempts by us moderns to recapture its past.

Chapter One

THE AGNOSTIC CONUNDRUM

> *The last English writer who professed to defend Christianity with weapons drawn from wide and genuine philosophical knowledge was Dean Mansel. The whole substance of his argument was simply and solely the assertion of the first principles of Agnosticism.*
>
> LESLIE STEPHEN

The Bampton Lectures had bored the English public ever since their institution in 1780 as a forum for the most traditional sort of Christian apologetics. However, in 1858 the lecturer's eloquence, wit, and brilliant powers of analysis attracted to St. Mary's the largest congregation since the days of John Henry Newman. The lecturer was Henry Longueville Mansel (1820–1871), a Tory, a High Church Anglican, at that time Reader in Moral and Metaphysical Philosophy at Magdalen College, and his lectures on *The Limits of Religious Thought* were a sensation.[1] The *Times* reported that "Sunday after Sunday, during the whole series, in spite of the natural craving for variety, and some almost tropical weather, there flocked to St. Mary's a large and continually increasing crowd of hearers, to listen to discourses on the Absolute and the Infinite, which they confessedly could not comprehend."[2]

Although the topic of the lectures was esoteric and philosophically complex, Mansel's hearers could grasp just enough of his meaning to know that his ingenious arguments were radically new and that they were considered by orthodox Christian leaders to be powerful ammunition for the war on unbelief. Mansel told his audience that the findings of German biblical criticism, French Positivism, and English geological science were unacceptable if they came into conflict with the Holy Scriptures. In defending the doctrine of biblical infallibility he did not differ from his predecessors who had undertaken the Bampton lectureship—it was how he argued his position which struck his listeners as novel and exciting. Mansel seemed to defend the most ancient form of orthodoxy through the use of the most modern weapons drawn from the theologically suspect philosophy of Immanuel Kant. Since man is a finite being with a conditioned consciousness, Mansel argued, his capacity for knowledge has definite limits. Both God and the transcendental world are beyond these limits and thus are unknowable. Therefore, man is in no position to criticize the Bible because it represents a communication from an inscrutable being (God). Only he who is omniscient can presume to evaluate the Scriptures. The infallibility of the Bible in all matters cannot be questioned.

The Old Saw of Agnosticism

In view of Mansel's unimpeachable Christian piety, it does not seem possible that the English biologist Thomas Henry Huxley (1825–1895), the man who coined the term *agnostic* to describe his own position, would enthusiastically recommend the Bampton Lectures to his friends. Yet Charles Lyell, the famous geologist, relates in a letter of 1859 the following description of Huxley's rather high opinion of Mansel's *Limits of Religious Thought.* "A friend of mine, Huxley, who will soon take rank as one of the first naturalists we have ever produced, begged me to read these sermons as first rate, 'although, regarding the author as a Churchman, you will probably compare him, as I did, to the drunken fellow in Hogarth's Contested Election, who is sawing through the signpost of the other party's public-house, forgetting that he is sitting at the outer end of it. But read them as a piece of clear and unanswerable reasoning.'"[3]

The picture referred to by Huxley is part of a series of paintings by William Hogarth (1697–1764) entitled *An Election: Four Pictures.* These four satirical pictures, based on the Oxfordshire election of 1754, symbolized, for Hogarth, England in the dark years from 1755 to 1757, when the folly of politicians had destroyed her military power and moral strength.[4] The second painting, *Canvassing for Votes,* was completed in 1757. Here, in front of a quaint village inn, are smiling politicians asking for and buying votes. However, behind this tranquil scene lurks an image of brutal violence, for in the distance a throng of men are trying to tear down a building, whose owner defends it by firing upon the unruly mob. Hogarth's concern with the evils of political corruption and the rising power of the mob are clearly reflected in *Canvassing for Votes.* A third theme is also presented therein—the absurdity of factionalism in politics. Above the seething crowd of men in the background, perched precariously atop a signpost, is a figure sawing down the sign of a public house that supports the party he opposes. The man has probably had a bit too much to drink, because he seems to

William Hogarth, "The Election, Plate II: Canvassing the Votes"

be unaware that if he succeeds in his task he too will fall. Hogarth is pointing out that a vicious war between political factions is ultimately self-defeating, for cutting down the opponent would destroy the whole political system, including one's own party.

Huxley took a tiny detail from Hogarth's painting in order to articulate his reaction to Mansel's Bampton Lectures. Originally set within a political context by Hogarth, the metaphor of mistakenly causing one's own downfall in sawing through the enemy's signpost is placed by Huxley within a theological context. The parties vying for power were no longer Whigs and Tories, but believers and unbelievers, Christians and scientific naturalists. In Huxley's hands the motley mob in the distance now becomes orthodox Christian theologians led by Mansel, the fellow up on the signpost. Mansel was "drunk," in Huxley's opinion, because, by attempting to cut down unbelievers using a saw fashioned from Kantian metal, Mansel had unwittingly destroyed the foundations of traditional Christian theology.

But Huxley's reaction to the Bampton Lectures indicates that Mansel not only worked against himself by undermining his own position but also supplied unbelievers with arguments that enabled them to construct a new form of scepticism, later labeled by Huxley as agnosticism. Mansel's reasoning was so "clear and unanswerable" that it became the essence of the agnostic viewpoint. As strange as it may seem, agnosticism owed a great debt to an eminent High Church Anglican, and Huxley was perversely fond of pointing out the similarity between Mansel's position and his own. In an article of 1895, Huxley remembers that when he came across *The Limits of Religious Thought*, he said to himself, "'Connu!'; and the thrill of pleasure with which I discovered that, in the matter of Agnosticism (not yet so christened), I was as orthodox as a dignitary of the Church, who might any day be made a bishop, may be left to the imagination."[5]

Huxley's use of the metaphor drawn from Hogarth's painting is actually doubly ironic. From Huxley's point of view it is ironic to come across a Christian theologian who, in holding to the notion of the limits of knowledge, is self-destructive and supplies unbelievers with powerful arguments. However, a second irony is concealed in the adoption by Huxley and the agnostics of Mansel's way of conceiving the limits of knowledge. A philosophical justification of the axioms upon which science must be based could not be undertaken by the agnostics if they restricted knowledge to the same degree as did Mansel. Andrew S. Pringle-Pattison (1856–1931), a Scottish philosopher, once compared the argument of Mansel's Bampton Lectures to "edged tools," saying that their inventor might escape evil but that "the next to handle them will surely cut their fingers."[6] Scepticism in general has been found to be a dangerous weapon, for it can often be two-edged. The variety of scepticism embraced by Mansel and Huxley was particularly potent, and they both "cut their fingers" on the blade of the saw they used to bring down their enemies. In the spirit of Hogarth's satiric art, we can visualize a drunken Huxley sitting right in front of an equally inebriated Mansel on that signpost, watching Mansel fall and then sawing off his own section of wood. For whereas Huxley was correct in saying that Mansel undermined orthodox Christianity, it is equally evident that Huxley undercut the certainty of science.

The double irony arising from Huxley's use of Hogarth's painting and the implications of that irony for Victorian unbelief will be the main theme to be explored in this study of the origins of agnosticism. The importance of Huxley's reaction to *The Limits of Religious Thought* is too often overlooked and has not received full treatment in studies of agnostic thought. A good reason for this lack of attention to the connection between Mansel and Huxley is that a rather loose definition of agnosticism obscures the true origins of this unique form of scepticism.

Huxley the Neologist

In 1882 a letter appeared in *Notes and Queries* asking for the date of the earliest use of the word *agnosticism*. Use of the term had become common and widespread enough by the early 1880s that people were becoming interested in its origin. After several correspondents pointed to publications in which the word appeared in 1876 and 1874, James A. H. Murray, the noted philologist, credited Huxley with coining the term in 1869.[7] When the *Oxford English Dictionary*, edited by Murray, was first published in 1884, invention of the word was again attributed to Huxley.[8]

Thomas Henry Huxley was born into an impoverished, lower-middle-class family. He studied medicine and then entered the Royal Navy medical service in 1846. Just as Darwin had received important scientific training through his experiences on a long sea voyage aboard the *Beagle*, Huxley's career gathered momentum between 1846 and 1850, while he was doing research as assistant surgeon and naturalist aboard the HMS *Rattlesnake*. Moving into the realm of biological and zoological research, Huxley was appointed lecturer at the Royal School of Mines in 1854, and then later he moved up to professor. Huxley subsequently held professorships at the Royal Institution and the Royal College of Surgeons in addition to the deanship of the Normal School of Science at South Kensington (now known as the Imperial College of Science and Technology). He enjoyed a long and distinguished career as one of Victorian England's greatest scientists and popularizers of science due to his unwearying efforts in public lecture halls, in the pages of fashionable periodicals, and in important government committees. Huxley was also notorious for his vigorous defense of evolutionary theory, which won him the title "Darwin's bulldog."

It was due to the respect accorded Huxley as one of the foremost scientists of the age that he was asked to join the Metaphysical Society, and it is significant that Huxley coined the term *agnosticism* in response to issues raised by the early meetings of this remarkable club. During its existence from 1869 to 1880, the members of the Metaphysical Society met nine times a year in London to hear prepared papers and discuss ultimate philosophical and religious questions. Among the membership were many of the major English thinkers of the time. Orthodox Christians such as Archbishop Manning, R. W. Church, W. E. Gladstone, and Connop Thirlwall were part of the society, as were the liberal-minded A. P. Stanley and F. D. Maurice. Men more left of center, but still within the pale of Christianity, such as W. R. Greg, R. H. Hutton, and James Martineau, were not averse to joining. Besides Huxley, other unbelievers, including J. A. Froude, Frederic Harrison,

Thomas Henry Huxley
A Photograph by Mayall, 1893

and John Morley, found a place within the society. They were joined by W. K. Clifford in 1874 and Leslie Stephen in 1878. As Huxley once remarked, "Every variety of philosophical and theological opinion was represented there, and expressed itself with entire openness" (*SCT,* 239).

In the company of his illustrious Metaphysical Society colleagues, Huxley began to feel somewhat embarrassed that he had no definite term to describe his philosophical position other than the rather vague *freethinker.* He rejected *atheist, theist, pantheist, materialist, idealist,* and *Christian* because those who were known by these appellations "were quite sure they had attained a certain 'gnosis,'—had, more or less successfully, solved the problem of existence; while I was quite sure I had not, and had a pretty strong conviction that the problem was insoluble." Challenged and attacked by the best minds in Victorian England, most of whom were "*-ists* of one sort or another," Huxley was forced to invent what he "conceived to be the appropriate title of 'agnostic,'" and, as he wrote, "I took the earliest opportunity of parading it at our Society" (*SCT*, 238–39).

Scholars have accepted without question Huxley's assertion in this section of his famous essay "Agnosticism" (1889, in *SCT*) that he coined the term in reaction to the Metaphysical Society meetings. However, there has been some confusion as to the source from which Huxley derived the word. The confusion began when Murray accepted Hutton's account in a letter dated 13 March 1881. Theologian, journalist, and man of letters, Richard Holt Hutton (1826–1897) was editor of the *Spectator* and a member of the Metaphysical Society. As the self-appointed chronicler of the agnostic movement, Hutton supplied the readers of the *Spectator* with a steady stream of articles on Huxley, and he is even credited with being the first to publish Huxley's coinages *agnostic* and *agnosticism*.[9] Huxley recalled that when he showed off his new label at the Metaphysical Society "the term took; and when the *Spectator* had stood godfather to it, any suspicion in the minds of respectable people, that a knowledge of its parentage might have awakened was, of course, completely lulled" (*SCT*, 239). Even Huxley admitted Hutton's key role in popularizing the term, and if not for Hutton, *agnosticism* might have remained part of the private language of a small circle of Victorian intellectuals. According to Hutton, Huxley had suggested *agnostic* at a party held at Knowles's home in 1869, before the formation of the Metaphysical Society, and had taken it from St. Paul's mention of the altar to the "Unknown God" in Acts 17:23.[10] Most scholars since then have trusted Murray's confidence in Hutton's letter.[11]

However, in the same section of "Agnosticism" which we have been examining, Huxley presents a very different explanation of the etymological source of the term he coined.[12] Huxley asserts that *agnostic* "came into my head as suggestively antithetic to the 'gnostic' of Church history, who professed to know so much about the very things of which I was ignorant" (*SCT*, 239). In an unpublished letter of 10 December 1889, which has not previously been cited by scholars in discussions of this issue, Huxley explicitly denied that *agnostic* was derived from Acts:

> The term "agnostic" was not suggested by the paragraph in the Acts of the Apostles in which Paul speaks of an inscription to the unknown God *(agnostic theo)*. It is obvious that the author of this inscription was a theist—I may say an anxious theist—who desired not to offend any God not known to him by ignoring the existence of such a deity.

> The person who erected the altar was therefore in the same position as those philosophers who in modern times have brought about the apotheosis of ignorance under the name of the "Absolute" or its equivalent. "Agnostic" came into my mind as a fit antithesis to gnostic—the gnostics being those ancient heretics who professed to know most about those very things of which I am quite sure I know nothing—Agnostic therefore in the sense of a philosophical system is senseless: its import lies in being a confession of ignorance—a warning set up against philosophical and theological phantasms which was never more needed than at the present time when the ghost of the "Absolute" slain by my masters Hume and Hamilton is making its appearance in broad daylight. (ICST-HP 30:152–53)

Determining whence Huxley derived *agnostic* is not simply a matter that should concern etymologists, because, as the above quote indicates, we can determine the general thrust of the term from its linguistic origin.[13] First, Huxley clearly tied an epistemological element to agnosticism and intended it to denote a profession of ignorance. Second, he saw agnosticism as the opposite of gnosticism.[14] The Gnostics were a sect existing both within and without Christianity and Judaism in the first three centuries A.D. Claiming to possess superior knowledge derived from secret revelations, the Gnostics were eventually driven out of the Christian Church. In calling himself an a-gnostic Huxley was underlining the orthodox quality of his position. The early Church was a-gnostic in proclaiming gnosticism heretical, and Huxley was siding with the early Christian leaders. If Victorian Christians were unwilling to accept the validity of Huxley's agnosticism, then, Huxley was cleverly implying, perhaps nineteenth-century Christianity was a new gnostic sect dogmatically claiming possession of higher knowledge.[15] Some Christian thinkers admitted that agnosticism was a somewhat justifiable response to the wild extravagances of theology. "For much of the Agnosticism of the age," James Martineau declared, "the Gnosticism of theologians is undeniably responsible."[16] It is only upon perceiving that Hutton's account of the linguistic origin of agnosticism is incorrect that the "orthodox" meaning behind Huxley's new word can be appreciated. However, for a more specific definition of agnosticism we must examine sections in Huxley's work that deal with the essence of the agnostic position.

Defining the Term Agnosticism

If we turn to scholarly literature on religious thought it is not entirely clear what criteria we should use to decide who is, and who is not, an agnostic. An astonishing number of thinkers besides Huxley have been referred to explicitly as agnostics or as espousers of agnosticism by theologians, historians, and philosophers. The list includes Heraclitus, Protagoras, Gorgias, Socrates, Carneades, Sextus Empiricus, Maimonides, Occam, Peter D'Ailly, Luther, Henry Cornelius Agrippa of Nettesheim, Faustus Socinus, Montaigne, Peter Charron, Pascal, Daniel Peter Huet, Pierre Bayle, Archbishop William King, Bishop Peter Browne, John Hutchinson, Hume, Kant, Goethe, Schleiermacher, James Mill, Lamennais, Sir William Hamilton, Carlyle, Comte, J. S. Mill, George Jacob Holyoake, Arthur Hugh Clough, George Eliot, Henry Longueville Mansel, John Tyndall, Herbert Spencer, Matthew Arnold, Albrecht Ritschl, George Meredith, Leslie Stephen, Samuel Butler, Algernon Charles Swinburne, Henry Sidgwick, Auguste Sabatier, William James, William Kingdon Clifford, Francis Herbert Bradley, Mrs. Humphry Ward, Alfred North Whitehead, Bertrand Russell, Martin Buber, Karl Jaspers, and Gabriel Marcel.[17] If thinkers as disparate as Socrates, Luther, Goethe, and Russell can be labeled agnostics, then it can fairly be asked if the term *agnostic* might not stand in need of radical redefinition.

Huxley's definition of the term he coined can help us begin to trim down this grossly inflated catalogue of names. Although there were times when Huxley himself, carried away by the heat of controversy and his own polemical skill, used the word *agnosticism* rather loosely, it is fairly clear what he intended. In those key sections of Huxley's work where he deals with his conception of agnosticism, two elements will always be found: a discussion of Kant or a thinker profoundly influenced by Kant, and an elaboration of Kant's notion of the limits of knowledge. For example, in *Hume* (1878) Huxley presented one of his earliest uses of the term *agnosticism* in print within the context of a discussion of Hume and Kant. "If, in thus conceiving the object and the limitations of philosophy," Huxley wrote, "Hume shows himself the spiritual child and continuator of the work of Locke, he appears no less plainly as the parent of Kant and as the protagonist of that more modern way of thinking, which has been called 'agnosticism,' from its profession of an incapacity to discover the indispensable conditions of either positive or negative knowledge." Although the details of Kant's critical philosophy differ from those of Hume, "they coincide with them in their main result, which is the limitation of all knowledge of reality to the world of phenomena revealed to us by experience." In the essay "Agnosticism" Huxley recalled how he steadily gravitated toward the conclusions of Hume and Kant, as they were summarized in a quotation from *The Critique of Pure Reason* which presented reason as an organ whose proper use is to limit knowledge. This section is strategically placed just prior to Huxley's story of how he coined the term when confronted by his Metaphysical Society colleagues.[18]

Huxley therefore conceived of agnosticism as a theory that restricted knowledge to the phenomenal realm and that was based on Kant's notion that the human mind is subject to inherent limitations. The essence of the agnostic argument was epistemological.[19] Although often directed at claims to certain knowledge of God, agnosticism could as easily say that claims to knowledge of self or an external world composed of matter are baseless. Any object that could be termed part of the transcendental or noumenal world was considered to be beyond the limits of human knowledge.

On the basis of defining agnosticism as a species of scepticism built upon Kantian principles, Huxley, Spencer, Tyndall, Stephen, and Clifford are bona fide agnostics. Herbert Spencer (1820–1903), the great "synthetic philosopher," put forward a full-blown program of agnosticism in 1860 and later accepted the term coined by Huxley as an accurate designation for his religious position. Leslie Stephen (1832–1904), known for his work as a philosopher, critic, and biographer (he edited the *Dictionary of National Biography*), was a self-professed agnostic. John Tyndall (1820–1893), professor of natural philosophy at the Royal Institution, and William Kingdon Clifford (1845–1879), professor in applied mathematics at University College, London, did not refer to themselves as agnostics in their published works. But both Tyndall and Clifford, along with Spencer, Stephen, and Huxley, presented the Victorian public with controversial essays and books articulating the agnostic position. They will therefore be the main focus of this study of agnosticism.[20]

From time to time I will examine the ideas and works of Victorian agnostics who played a less influential role in constructing the agnostic viewpoint. This list includes men such as Charles Darwin, whose spiritual odyssey from orthodox Christianity to agnosticism seems of immense significance in light of his discovery of the theory of natural selection. However, Darwin never published anything of substance on his agnosticism, just bits and pieces concerning his religious thought scattered throughout his writings and a brief section in his bowdlerized *Autobiography*, which appeared in 1887, too late to be considered decisive for the development of Victorian agnosticism. John Morley (1838–1923), editor of the *Fortnightly Review* from 1867 to 1882 before he went on to focus his energies on politics as a devoted Liberal, is another example of one who wrote little on his agnosticism for his contemporaries. Other agnostics who were hostile toward established Christianity in England in their publications, but who played a minor role in the formulation of agnostic theory, were Francis Galton (1822–1911), founder of eugenics, and Edward Clodd (1840–1930), banker, author, and Huxley's biographer. I will also discuss the work of lesser-known agnostics, such as Samuel Laing, Frederick James Gould, and Richard Bithell.

Atheism, Agnosticism, Theism

There are three more thinkers who deserve to be classed as authentic agnostics if we adhere to the definition that has been presented. Kant, Hamilton, and Mansel all would qualify. This follows from a definition of agnosticism which stresses its epistemological nature rather than its apparent antireligious bias. What is essential about agnosticism, and what Kant, Hamilton, and Mansel all share with Huxley, Tyndall, Clifford, Stephen, and Spencer, is the belief that there are inherent and constitutive limits of human cognition. In addition, they all would agree that we are ignorant of God's true nature since he is a transcendental entity and therefore outside the limits of human knowledge.[21]

If we include Kant, Hamilton, and Mansel as agnostics, then we are confronted with the possibility of a species of agnosticism which is Christian, theistic, and religious, a thought that jars the modern sensibility. We are usually accustomed to conceiving of agnosticism as, to move from the particular to the general, hostile toward Christianity, atheistical, and certainly irreligious. Yet even some of the self-professed agnostics do not fit into these categories of unbelief.

Writing just after the turn of the century, Benn observed that agnosticism excluded "Christian belief."[22] Yet it is not at all certain that the theological doctrines attacked by the Victorian agnostics—for example, the dogma of biblical infallibility, the notion of heaven and hell, and the belief in miracles—are necessary to the existence of Christianity. They may have been seen as essential by the Christians of Huxley's era, but Christianity has been transformed many times throughout history, and tenets considered as orthodox during one period have been jettisoned in other times. Far more common is the stronger charge that agnosticism is really atheistic, which implies that agnosticism is also anti-Christian. Henry Wace, later Dean of Canterbury, sounded a theme in "On Agnosticism" (1888) which was repeated by Huxley's contemporaries and later by twentieth-century thinkers. Wace charged that the adoption of the term *agnostic* was only "an attempt to shift the issue," "a mere evasion," for the agnostic's "real name is an older one—he is an Infidel, that is to say, an unbeliever."[23]

Victorian orthodoxy has received support for its claim that agnosticism was used as a disguise for a genuine atheism from an unexpected source—two important Marxists, Engels and Lenin. In his introduction (1892) to *Socialism Utopian and Scientific*, Engels referred to agnosticism as "'shamefaced' materialism," linking materialism with a denial of the existence of a supreme being. Lenin later reiterated this point in *Materialism and Empirio-criticism* (1908). Huxley's "agnosticism serves as a fig-leaf for materialism," Lenin joked, while simultaneously lampooning the Englishman's prudish distaste for materialism as something to be embarrassed about, like one's genitals.[24] The Marxists, who saw in the agnostics inconsistent atheists, attacked Huxley and his ilk for not going far enough, while the Victorian Christians rejected the agnostics for going too far from an acceptable orthodox position.

Despite the claims of unsympathetic contemporaries, the agnostics did not always hold to an atheistic position inimical to Christian theism. Tyndall maintained publicly in 1870, in "Scientific Use of the Imagination," that evolutionists "have as little fellowship with the atheist who says there is no God, as with the theist who professes to know the mind of God" (*FS* 2:134). After delivering his "Belfast Address," Tyndall had great difficulty disabusing his critics of the notion that he was an atheist. In an unpublished letter of 7 September 1874, he wrote: "The people that raise this uncandid outcry are not worthy of contradiction. They would roast me, but the time of roasting is happily gone by. You *are* correct in saying that *I am not an Atheist*. Though I am far from accepting their crude notions of the Power that rules the Universe."[25] Huxley and Stephen also repeatedly denied the accusation of atheism.[26] As agnostics, they believed that humans were incapable of gaining certain knowledge of God, but they agreed that from this epistemological position it followed that positive denial of God's existence was out of the question.

Many who do agree that there is a genuine difference between atheists and agnostics tend to set up a schema that places agnosticism midway between atheism and theism. The agnostic is one who rejects theism but is not quite an atheist, or one who suspends judgment concerning the existence of God. Similarly, agnosticism is portrayed as a neutral position.[27] Besides excluding all theists such as Spencer and Tyndall from consideration as agnostics, this definition tends to be so hazy that it becomes the justification for including almost all doubters, many of whom do not adhere to the typical agnostic theory of knowledge.

But by far the most misguided approach to agnosticism is one that perceives it to be antireligious, for this undoubtedly implies that agnosticism is atheistic and anti-Christian.[28] Preaching to his fellow Christians in 1884, the latitudinarian Reverend A. W. Momerie, Professor of

Logic and Metaphysics at King's College, London, declared that if agnosticism "be true, faith is a mistake; prayer is a mockery; to hope for immortality is as unreasonable as to hope for wings. Nothing worth calling a religion . . . can ever be founded upon an agnostic basis."[29] However, Hutton shrewdly tagged the agnostics "the adorers of Inscrutability," and pointed out that they provided themselves with "an equivalent for religion."[30]

A Sceptical Look at the Sceptical Tradition and Flint's Agnosticism

Just as agnosticism is often confused with atheism, it has also been conflated with other forms of unbelief, whether they be modern empiricism, materialism, positivism, or the development of pre-nineteenth-century scepticism.[31] To Robert Flint (1838–1910) the two words *sceptic* and *agnostic* were "about as nearly synonymous as any two words can be expected to be which refer to any comprehensive or complex phenomenon" and hence he concluded that "'sceptic' and 'scepticism', employed in their universally recognized and only philosophical signification would have served Professor Huxley just as well."[32] Flint's *Agnosticism*, published in 1903 but delivered as a set of lectures during the late eighties, is one of the best major studies of agnosticism to come out of the Victorian period.[33] At first, Flint tended to his flock as minister of the East Church, Aberdeen (1859–1862), and of Kilconquhan, Fife (1862–1864). But in 1864 he was elected to the chair of moral philosophy at St. Andrews University, and in 1876 he moved to the divinity chair of Edinburgh University. Flint was a liberal Christian who believed that Christianity derived its main strength from the ability of human beings to perceive the workings of God in history and in their own lives. It was this type of "religious knowledge" that Flint looked to as a counteracting force to agnosticism not only in his own day but in ages past as well.

Flint's *Agnosticism* is undeservedly neglected these days, because it is a thoughtful and perceptive book. His familiarity with nineteenth-century German and French thought, as well as his command of the whole tradition of scepticism in European thought, is impressive, and he is especially attentive to the epistemological dimension of agnosticism. However, Flint's desire to defend the Christian perspective is apparent in his whole approach to the history of agnosticism.

Flint divides the history of agnosticism into three periods, the Oriental, the Classical, and the Modern. Since he identifies agnosticism with scepticism, he is able to find proto-Huxleyites among Oriental thinkers, Greek sceptics, and Medieval nominalists. But although Flint discusses how these pre-nineteenth-century sceptics doubted the human ability to obtain certain knowledge in a variety of areas, he fails to demonstrate that, like the genuine agnostics, their delineation of the limits of knowledge was based on their investigation of the inherent structure of the mind. Flint admits that the first period of agnosticism, the Oriental, was "only of a rudimentary character," presenting us with "approximations to agnosticism, not with distinct forms of it" (79). The question of the limits of human knowledge, Flint concedes, was not specially discussed or distinctly raised. Even the first phase of modern agnosticism, from the beginning of the sixteenth century up to Hume, Flint sees as being "considerably different from the agnosticism of Hume and Kant, and of our contemporaries." Characterized by "imperfect development," since it "did not rest on any searching or comprehensive criticism of the powers of the human intellect," sixteenth- and seventeenth-century European scepticism was "mainly the expression of an exaggerated depreciation of knowledge or of a despair of acquiring knowledge" (100). At the beginning of the book Flint defined agnosticism as "the theory of the nature and limits of human intelligence which questions either the certainty of all knowledge and the veracity of every mental power, or the certainty of some particular kind of knowledge and the veracity of some particular mental power or powers" on the grounds "that the human mind is inherently and constitutionally incapable of knowing" (21). Therefore his attempt to include all sceptics prior to Kant as agnostics is inconsistent with his own definition.

Flint's tendency to overemphasize the line of continuity from pre-nineteenth-century scepticism to agnosticism is also questionable if we turn to a brief comparison of the different varieties of unbelief in European thought. Originating in ancient Greek thought, scepticism as a philosophical view was developed by the Academic sceptics into the position that no knowledge is possible and, even further, by the Pyrrhonian sceptics, who claimed that the Academics went too far in even making this statement. The Pyrrhonians believed that a suspension of judgment on all matters concerning knowledge was the only reasonable attitude.[34]

Both sceptical positions sunk into obscurity after the Hellenic age, until the Pyrrhonian view was revived in the sixteenth century due to the discovery of hitherto neglected manuscripts of Sextus Empiricus's writings. Popkin has shown that the intellectual crisis engendered by the Reformation led to the application of Pyrrhonian arguments to the problems of the day by sixteenth- and seventeenth-century thinkers. If we contrast Pyrrhonism to agnosticism we find that the arguments of the Greek sceptics are far more extreme and that they purposely undermine natural science, an aim the agnostics obviously repudiate. The agnostics were doubtful only about certain areas of knowledge, those that had to do with that transcendental realm beyond the limits of knowledge. What differentiates the agnostics from the sceptics dealt with by Popkin in *The History of Scepticism* is that they draw their arguments from Kant, not Pyrrho.

Among his menagerie of agnostics Flint included Bayle and Hume, allowing them to speak for the Enlightenment sceptics. Flint, then, believed that a genuine agnosticism existed in the eighteenth century and that Huxley and his agnostic colleagues were successors to the Enlightenment philosophes. To be sure, the agnostics looked upon a number of eighteenth-century philosophers as kindred souls, and they published essays and books devoted to rehabilitating their fallen reputations. Huxley's *Hume* (1878), Stephen's *History of English Thought in the Eighteenth Century* (1876), and Morley's studies of Burke, Voltaire, Rousseau, Diderot, and other Enlightenment figures were aimed at combating the feelings of horror that English intellectuals since the time of the Romantics had experienced when they looked back on the eighteenth century and its climax in 1789.[35] The agnostics saw in the writings of the philosophes a number of themes to be applauded—the refusal to look back to the ancients, a rejection of the Middle Ages as a time of superstition and oppression, and the attempt to create educational schemes that instill tolerance in individuals in order to work toward a "heavenly city" that would have no prejudice, ignorance, or unjust government. Similar as well is the animosity toward the Christianity of the day, the repudiation of revealed religion, the rabid anticlericalism, and the call for a purified Church. Perhaps more striking for our purposes is the parallel between the essentially religious foundations of Enlightenment thought and the significant Christian element in agnosticism.[36] Although tending to reduce religion to ethics and emotion, both forms of unbelief attempted to update religion by presenting a new faith that took into account the vast changes experienced by Europeans.

But there were important differences between the philosophes and their spiritual brothers of the nineteenth century. The philosophes were aristocrats who never talked about atheism in front of the servants, whereas the agnostics were more democratic in their belief that it was their duty (and to their advantage politically) to be outspoken in their public attack on the Christianity of the day. In terms of the actual content of its unbelief, the Enlightenment was far more negative and destructive, and hence less effective, than the agnosticism of the following century. Huxley remarked that his agnosticism differed from "its predecessor in the eighteenth century, in that it builds up, as well as pulls down." Voltaire's "scoffing doubt" was to be avoided as an evil in the same class as Christian bigotry.[37] The fatal weakness of Enlightenment intellectuals, according to Huxley, was their a priori philosophizing, which was unable to provide a "permanent resting-place for the spirit of scientific inquiry" (*SCT,* 18). Men such as Voltaire were not truly scientific in their overemphasis on natural religion. The God "proved" by Newtonian science and observed by all thinking men unaided by revelation was merely a rational construction of the intellect and not an empirically verified fact.[38] Huxley and the agnostics were more aware of the limits of reason and, unlike the philosophes, used epistemological arguments to attack the traditional Christian notion of God directly. Whereas eighteenth-century belief was based on a sensationalist theory of knowledge, the agnostics benefited from Kant's more subtle approach to epistemology through the structure of the mind. Even Hume, the Enlightenment thinker most often referred to as an agnostic, did not share the distinctive Kantian feature of agnostic thought.[39] And it was this more sophisticated epistemology, along with other advantages derived from the distinctiveness of nineteenth-century unbelief, that made the agnostic attack on traditional religion far more devastating than the Enlightenment attempt to *écrasez l'infâme.*

The tendency of Flint, writing from a Christian background, to assimilate agnosticism to scepticism is partly a result of his inability to see in agnosticism anything but irreligiousness. Wace's view, that all agnosticism is atheism, is echoed in the philosophical realm by Flint's position that all agnosticism equals scepticism. There is also an advantage to viewing agnosticism as identical to previous forms of scepticism, and Flint was not slow to utilize it. It was possible for him to critique modern agnosticism by attacking ancient scepticism. It was no doubt comforting to Flint and his readers that he could claim that Victorian agnosticism was nothing new and that Christian theologians had overcome this challenge to the faith before.

However, agnosticism was a unique phenomenon of unbelief that was more potent than any previous form of scepticism. This was true not only in terms of numbers of people who were profoundly affected but also from the viewpoint of its level of philosophical sophistication. The widespread popularity of agnostic ideas is tied up with a recognition that the development of agnosticism is grounded in the historical circumstances that molded the Victorian ethos. The superior cogency of agnostic arguments points to the tremendous raw energy the agnostics gained by tapping into Kant's powerful approach to epistemology in *The Critique of Pure Reason.* Although Flint believed that "the agnosticism of the present day flows directly from Hume and Kant" and that recent agnosticism owed to Kant "the larger part of what has given it plausibility and attractiveness, . . . very much of all that constitutes the superiority of recent agnosticism over earlier agnosticism," he denied that agnosticism should be conceived solely as being of modern growth. Likewise, he rejected the view that Kant was a revolutionary thinker who presented a radically new element in the development of scepticism (55–57, 117, 189).

In fact, Flint was highly critical of Kant, as well as others like Hamilton and Mansel, who adopted the agnostic position in order to defend a religious position. Flint was quite perceptive in recognizing that Mansel's brand of agnosticism destroys the philosophical justification for science while Huxley's scientific agnosticism is fatal to orthodox religion. Yet he saw no possibility of either a self-consistent agnosticism or an agnosticism that does not bear the seeds of its own destruction. A partial or modified agnosticism (i.e., one claiming to restrict its doubts to one type of knowledge) must, according to Flint, carry with it a demand to be put into its completed form, absolute or total agnosticism. For in destroying the credit of one department of knowledge the partial agnostic must hold, in order to be consistent, that the same argument is valid against all other departments of knowledge (193). To Flint this was no less true of religious agnostics like Kant and Mansel, who can only be hurtful to religion despite their sincere intentions. Although Flint was quite right that the agnosticism of Mansel and Huxley was ultimately self-defeating and inconsistent, he did not recognize the validity and inner integrity of Kant's agnosticism, which preserved both science and religion. Flint's insensitivity to the distinctiveness of Kant's position seriously mars what is otherwise an important work.

The Unique Place of Agnosticism in Nineteenth-Century European Unbelief

As Huxley groped throughout the late 1860s for a way to articulate his new position, he was faced with an almost insurmountable difficulty. How could he prevent his standpoint from being confused with Positivism, materialism, or empiricism, movements of thought which were far narrower in their meaning during the nineteenth century? Positivism was chiefly understood to be the philosophy of Comte, materialism was identified with German thinkers such as Büchner, and empiricism was regarded as the school of J. S. Mill, so the agnostics required their own label to signify their disapproval of tenets held by these other types of unbelief.[40]

In 1869, the same year that Huxley coined the term agnostic, he lashed out at those who too readily hurled the epithet "Positivist" at innocent agnostics and scientists:

> It has been a periodical source of irritation to me to find M. Comte put forward as a representative of scientific thought; and to observe that writers whose philosophy had its legitimate parent in Hume, or in themselves, were labelled "Comtists" or "Positivists" by public writers, even in spite of vehement protests to the contrary. It has cost Mr. Mill hard rubbings to get that label off; and I watch Mr. Spencer, as one regards a good man struggling with adversity, still engaged in eluding its adhesiveness.[41]

An authentic Positivist in the 1860s was one who followed Comte in holding that the phenomena of human thought and of social life are continuous with the natural world and are thereby subject to the Law of the Three Stages, as well as susceptible to investigation through scientific methods, the relative validity of which were determined by Comte's classification of the sciences.[42] In addition to this body of doctrine regarding the nature and place of sociology as a science, the genuine Positivist had to digest an elaborate set of dogmas forming the articles of a Religion of Humanity.

Although the agnostics were attracted by the Positivist's stress on a scientific approach to social problems, the rejection of metaphysics, and the view of science as the ideal form of knowledge, nothing could induce them to swallow Positivism hook, line, and sinker, as had Richard Congreve, Edward Beesley, Frederic Harrison, and other English disciples of Comte. The agnostics were anxious to stress their differences from the English Positivists, and Huxley led the charge. In his essays "On the Physical Basis of Life" (1868, in *MR*) and "The Scientific Aspects of Positivism" (1869, in *Lay Sermons, Addresses, and Reviews*) Huxley subjected Comte to a devastating critique. It was Huxley's contention that Comte's positive philosophy contained nothing "of any scientific value" and that the spirit of modern science was founded by Hume and not Comte. Huxley went even further in undermining the Positivist claims to scientific authority in his famous remark that "Comte's philosophy, in practice might be compendiously described as Catholicism minus Christianity."[43] In Huxley's eyes, Positivism was more a pseudoreligion than a strictly scientific body of knowledge.

The other agnostics were delighted with Huxley's deflation of Comte's pretensions and with Huxley's attempt to distinguish Positivism from true science. Morley, then editor of the *Fortnightly Review*, which had published Huxley's essays with the attacks on Comte, wrote to Huxley: "I fully understand the vexation with which you have undergone the popular or archiepiscopal confusion about every scientifically minded person being a Comtist; and I hope your protest will do something to clear people's heads."[44] Tyndall remarked to Huxley in a letter dated 12 January 1869, "I was much amused by the *Leaders* remarks on your really just criticism of Comte."[45] Probably most gratified of all the agnostics was Spencer, whose violent dislike of Comte stemmed from the fact that Positivism offered a universal and scientifically based system of knowledge which competed with his own synthetic philosophy. Spencer later found himself embroiled in a controversy over the comparative advantages of agnosticism to Positivism with Frederic Harrison in the mid eighties.

An important area of disagreement between agnostics and Positivists which is rarely discussed centers on epistemological issues. Was it true that the agnostics learned their epistemology from Comte when they subscribed to his emphasis on seeking the laws of things or the invariable relations of succession and similarity, rather than attempting to discover the inner causes of phenomena?[46] Although the agnostics agreed with Comte that we can have no knowledge of anything but phenomena, Comte held to this position without investigating the human claim to knowledge. Huxley, like the other agnostics, insisted on the validity of a scientific psychology as the basis of a sound philosophy and took Comte to task for leaving epistemology and psychology off his table of sciences.[47]

But if Huxley did not accept the term *Positivism* as the correct designation for his position, then, his contemporaries asked themselves, what did he call himself? After having rejected Comte in the late 1860s, Huxley was faced with the problem of avoiding the name tag *materialist.* Indeed, Huxley's attempt in "On the Physical Basis of Life" to "prove the existence of a general uniformity in the character of the protoplasm, or physical basis, of life, in whatever group of living beings it may be studied" made it that much more difficult to escape the charge of materialism.[48] Huxley complained to Tyndall that his essay had been totally misunderstood. "The paper upon the Physical Basis of Life was intended by me to contain a simple statement of one of the greatest tendencies of modern biological thought, accompanied by a protest from the philosophical side against what is commonly called materialism. The result of my well-meant efforts I find to be, that I am generally credited with having invented 'protoplasm' in the interests of materialism." All this despite Huxley's declaration in the essay that "I, individually, am no materialist, but, on the contrary, believe materialism to involve grave philosophical error."[49]

Tyndall, Stephen, Clifford, and Spencer also had to contend with charges that they were materialists. Just as repugnant for many Victorians as Huxley's protoplasm was Tyndall's "Belfast Address" (1874), wherein he boldly discerned in matter "the promise and potency of all terrestrial life." However, Tyndall repudiated the title "materialist" as did Huxley. "People sometimes revile me for being 'a materialist,'" Tyndall recorded in his journal in 1872, "as if I as much as they, and in many cases a thousand times more than they, would not rejoice to see what they call the spirit liberated more than it now is from the dominion of matter." Stephen was equally adamant. "I have been told," Stephen declared, "as a matter of course, that I am a Materialist. I do not think that I am one in any fair sense of the word, but I willingly leave it to others to label me with such tickets as they please in the museum of monstrosities." What grated most on the agnostics' nerves was, as Spencer put it in a letter to Huxley, the way in which "you and I are dealt with after the ordinary fashion popular with the theologians, who practically say—'You *shall* be materialists whether you like it or not.'"[50] Most nineteenth-century critics did not take these denials of materialism seriously.[51]

During the nineteenth century, however, European materialism was limited almost exclusively to Germany in the form of the scientific materialism of men such as Vogt, Büchner, and Moleschott and the dialectical materialism of Marx and Engels. With philosophical positions held by these men agnosticism had little in common. The German scientific materialists developed their popular hodgepodge of atheism, anticlericalism, and reductionism during the 1840s. In rejecting the old German transcendentalist tradition and *Naturphilosophie*, the scientific materialists made extensive claims for the power of science to explain all phenomena. Where Büchner unashamedly appropriated the title of materialist as "a title of honour" and made the principle "No force without matter—no matter without force" the basis of his immensely popular *Kraft und Stoff* (1855), Huxley humbly confessed that he had "never been able to form the slightest conception of those 'forces' which the Materialists talk about." While Büchner and the scientific materialists directed their attacks on the German neo-Kantians, who stressed the limits of knowledge, Huxley believed that materialists transgressed these limits in their claim that everything is composed of matter in forms determined by the working of forces.[52] In his essay "Science and Morals" (1886), Huxley referred explicitly to *Kraft und Stoff* as espousing a "faith materialistic" and stated his reasons for "heartily disbelieving" Büchner's philosophy (*EE*, 129).

The agnostics were prepared to fight to the death to defend the right of scientists to remain strictly on the material level when analyzing physical phenomena, since materialistic terminology had proven in the past to help people control nature better than obscure spiritualistic terminology. But equally important to the agnostics was the recognition that the scientist erred who tried to convert his materialistic description of nature into an actual ontological doctrine.[53] Although Huxley and his agnostic colleagues could sympathize with the scientific materialists' belief in the importance of science and their stress on the eradication of ignorance and superstition and their emphasis on the need to banish supernatural causes from science, the disagreement of the two groups about the significance of Kant's epistemology led to a radical difference in the whole thrust of their respective views.[54]

Turning to dialectical materialism, we again find a basic discontinuity between this type of materialism and the agnostic position. I have already discussed Engels and Lenin's sarcastic references to agnosticism as half-hearted materialism. Engels was critical of the agnostics for postulating the existence of mysterious and ungraspable objects because he believed that science would eventually be able to analyze all things into their chemical elements. Lenin entirely agreed with Engels and traced agnosticism back to Hume and Kant. "Those who hold to the line of Kant or Hume," Lenin declared, "call us, the materialists, 'metaphysicians' because we recognize objective reality which is given us in experience, because we recognize an objective source of our sensations independent of man. We materialists follow Engels in calling the Kantians and Humeans *agnostics* because they deny objective reality as the source of our sensations."[55]

Forms of Unbelief "Made in England"

In addition to the confusion of agnosticism with Positivism and materialism, Huxley was confronted by the possibility of being saddled with the label "empiricist." Unlike the philosophies of Positivism and materialism, which had to be imported from across the channel, empiricism was a distinctly English intellectual tradition. Stephen's remark that "the critical movement initiated by Locke and culminating with Hume reflects the national character" has been reiterated by many scholars in their discussions of the English people's practical bent of mind, profound respect for facts, and emphasis on empirical experience.[56] The empiricist strain in English thought proved to be incredibly resilient, for despite the pervasive influence of early nineteenth-century Romanticism in the form of Coleridge, Carlyle, and the Oxford Movement, the native tradition lived on in philosophical radicalism and political economy, and gave birth to a second generation of Utilitarians, who restored the empiricist school to a position of dominance during mid century. John Stuart Mill's *System of Logic* (1843), according to Leslie Stephen, was "a kind of sacred book for students who claimed to be genuine Liberals." The *Logic* was used as a textbook at Oxford by the 1850s, and Mill's immense prestige continued well into the eighties.[57]

For those English thinkers who condemned the growth of agnosticism, Engels offered the dubious consolation that at least "these 'newfangled notions' are not of foreign origin, are not 'made in Germany,' like so many other articles of daily use, but are undoubtedly Old English, and that their British originators two hundred years ago went a good deal further than their descendants now dare to venture."[58] Engels's remark is a good example of the tendency to see in agnosticism the logical culmination of the English empiricist tradition.[59] J. S. Mill is usually referred to as an agnostic himself, as is his Utilitarian father, James.

However, Huxley's creation of the new term *agnosticism* was also intended to signify that his position was to be distinguished from empiricism, which in the 1860s meant being a disciple of J. S. Mill and Alexander Bain. Although the agnostics clearly were attracted to many of Mill's beliefs, such as the rejection of intuitionism, the emphasis on experience as the source of truth, and the main tenets of Utilitarianism, they were under no illusions as to the deficiencies of Mill's philosophy. While the agnostics used evolutionary theory to understand and explain almost every feature of the universe, the majority of Mill's work was undertaken either before the appearance of *The Origin of Species* (1859) or during the sixties, when the significance of Darwin's thought was not clear. Stephen voiced an important agnostic theme in his belief that Utilitarianism required "re-statement or reconstruction" in light of developments in evolutionary theory.[60] The agnostics recognized that it was hardly reasonable to have expected Mill to undertake this task of reformulating empiricism himself. Yet they could hold Mill responsible for the surprising reticence he displayed when confronting religious issues, especially in the posthumously published *Theism* (1874). Two years later, Stephen the plainspeaker criticized Mill for "a pathetic desire to find some remnant of truth in the ancient dogmas [which] breathes throughout its pages, and is allowed to exercise a distorting influence upon its conclusions."[61] Furthermore, Mill never held to the essential agnostic position that God is unknowable owing to the inherent limitations of the human mind. Mill, unlike Huxley, did not need a new label to describe his position. He already possessed a creed that he had discovered in his youth. He recalls in his

Autobiography (1873) that after reading Bentham he had "a creed, a doctrine, a philosophy; in one among the best senses of the word, a religion."[62]

Besides empiricism there were other English varieties of unbelief in the nineteenth century. During the late forties and fifties a small group of independent thinkers, including A. H. Clough (1819–1861), Francis W. Newman (1805–1897), Tennyson (1809–1892), and J. A. Froude (1818–1894), questioned the validity of traditional Christianity in their poems, novels, and spiritual autobiographies. These writers were not concerned with advancing epistemological arguments for the unknowability of God; it was institutional religion that repelled them.[63] Their faith in the authority of the Church had been shaken by the findings of German biblical criticism, moral objections to Christian doctrine, and the Oxford Movement's denunciation of the stagnation of Anglicanism, which drove them beyond the pale of orthodox Christianity rather than Romeward. The unbelievers of the forties and fifties rebuilt their faith on a reverence for the human spirit. In a review of F. W. Newman's *The Soul* (1849), Clough approved of Newman's rejection of a faith based on historical facts and doctrinal articles because "the abiding revelation is written, not on hard tablets of stone, legal, historic, or dogmatic, but on the fleshly tablets of the human heart and conscience."[64]

The original agnostics were not atheists, nor were they materialists or Positivists. Their stress on Kantian epistemology distinguishes them from any previous form of British unbeliever, including the empiricists or the unbelievers of the forties and fifties. However, the agnostics do belong to a distant group of English intellectuals who were hostile toward the Victorian Church. Agnosticism was an important component of the nineteenth-century movement known as "scientific naturalism."

Scientific naturalism was the English equivalent of the cult of science in vogue throughout Europe during the second half of the nineteenth century. During the ongoing debate on man's place in nature in Victorian England (a public discussion that encompassed all realms of thought) the scientific naturalists put forward new interpretations of man, nature, and society derived from the theories, methods, and categories of empirical science. This cluster of ideas and attitudes was naturalistic in the sense it would permit no recourse to causes not present in empirically observed nature, and it was scientific because nature was interpreted according to three major mid-century scientific theories, the atomic theory of matter, the conservation of energy, and evolution. The ringleaders of scientific naturalism were Huxley, Tyndall, Spencer, Clifford, Galton, Harrison, Morley, G. H. Lewes, Edward Tylor, John

Lubbock, E. Ray Lankester, Henry Maudsley, Stephen, Grant Allen, and Clodd.[65]

Even though scientific naturalism is a broader movement of thought than is agnosticism, examining the connection between the two will contribute valuable insights into the birth of Huxley's brainchild.[66] All general points regarding the ethos of scientific naturalism largely hold for the agnostics. Moreover, agnosticism shared a common social context with scientific naturalism. Conclusions drawn from a study of the social significance of scientific naturalism apply with equal force to agnosticism. The ideology of scientific naturalists became the apologetic tool of the Victorian middle class in its attempts to generate a new *Weltanschauung*, one appropriate in a competitive, urban, and industrial world, as a replacement for old philosophies and theologies suitable to a pastoral, agrarian, and aristocratic world.[67]

The efforts of scientific naturalists were resisted by the Church and the propertied classes, whose alliance was strengthened during the first half of the nineteenth century, when religious infidelity and politically dangerous ideas were seen as complementary. Christianity had assumed the role of defender of the social order in many European nations, wherein fortifying the coalition between throne and altar was the response to the threat of revolution in the wake of 1789. Stephen declared that Christianity had worked "itself so thoroughly into alliance with the conservative forces of society that it is no longer possible to separate the two interests. Its influence is rigorously dependent upon the strong conviction of the governing classes that the old creed is bound up with the old order" (*AA*, 364). The clash between social classes, therefore, had a political dimension, and scientific naturalists were by and large supporters of the radical wing of the Liberal party, at least to begin with, as "young Turks."

But despite the animosity of middle-class scientific naturalists toward the Church, we must not forget the factors of continuity which linked the two warring factions. As the nineteenth century wore on, the Victorian middle class increasingly began to feel that it was entering into a political partnership with the upper class and that it had a stake in the continued health of the social order. In the intellectual realm there existed subtle parallels with the old system under attack, not the least of which flourished in the area of religious thought.

In his *History of English Thought in the Eighteenth Century* (1876) Leslie Stephen made the following comment about the general tendency of ideas to persist in the ebb and flow of intellectual history:

> The most unflinching sceptic really carries with him far more than he knows of the old methods of conception. He inherits the ancient

> framework of theology, and, unable to find a place in it for his new doctrine, cuts away a large fragment to make room for the favourite dogma. To his contemporaries this sacrilegious act appears to be the most important; it is the mark by which they recognize his peculiar character; to observers at a distance it may appear that his conservatism is really more remarkable than his destructiveness. They wonder more that he should have retained so much than rejected so much. He follows the old method or retains the old conception, though he sees its futility for attaining the old ends. The discord is the result of an incomplete transformation of thought. He gives up hell, but he admits that hell is the only sanction for morality. (1:9)

Stephen's critique of "incomplete" scepticism, although aimed at the deists of the eighteenth century and perhaps even the Broad Churchmen of his own day whom he so despised, is applicable to his fellow agnostics. It also suggests a view of agnosticism which preserves both change and continuity in the development of Victorian thought. Stephen's contemporaries were impressed most by the agnostic attack on the Anglican Church. But Stephen also reminds us that what struck contemporaries as important differences between unbelievers and the orthodox should not blind us to the larger continuity between the old and new, and hence to the religious quality of agnosticism. Living in an age of unprecedented change and transition, the Victorian agnostics produced a form of thought which reflected the times. The Victorians were caught between two worlds, the medieval era and the slowly emerging modern age. It is no coincidence that the Victorian period was also the agnostic epoch *par excellence*. Only then did agnosticism crystalize into a widespread movement of thought which gloried in the sacredness of uncertainty.

The Missing Link

When Huxley extolled the virtues of Mansel's Bampton Lectures he was disclosing two facets of the ideological significance of agnosticism. First, he indicated that Mansel's arguments would be valuable aids in the attempt of scientific naturalists to discredit the authority of the Church. The notion of the limits of knowledge was a useful tool for revealing the gnostic pretensions of Christian thinkers and thereby questioning their ability to lead England into the brave new world of the future. Second, he revealed that those who transferred ideas from the lectures into scientific agnosticism shared some of the interests of the lecturer. Mansel desired to defend the Bible and constitutional authority as a bulwark against democracy and reform, while the agnostics were motivated by the more moderate concern of preserving social order within a bourgeois society.[68]

Although an understanding of the social context allows us to understand why the work of an orthodox Christian such as Mansel was attractive to the agnostics, it still leaves open the question of how Huxley and his colleagues altered the message of *The Limits of Religious Thought* to suit their own ends, and it fails to explain the consequences that arose for their articulation of the agnostic viewpoint. To deal with these issues we must probe as deeply as possible at the level of intellectual content.[69] This means discussing agnosticism in the context of the importation of German modes of thought into England after 1850 and their synthesis with English empiricism. German thought is usually seen as infiltrating England in a significant way during the seventies and eighties, once the fortunes of scientific naturalism had begun to wane, with the development of neo-Hegelianism. However, German ideas made their way into England decades earlier, and one of the mediums of transmission was the agnostics themselves. Although the ability to read German was rare in England during the third quarter of the nineteenth century, Huxley, Tyndall, Stephen, and Clifford were among the few who possessed the skill.

Mansel is the "missing link" in the history of the theory of agnostic descent.[70] It is through a study of his thought that we can perceive the connection between the species known as the Kantian tradition and the agnostic species. Historians have tended to deal with the decline in traditional religious faith and the corresponding rise of agnosticism by emphasizing the destructive effects of science and evolutionary theory, the impact of biblical criticism, and the role of the ethical revolt from orthodox Christianity.[71] However, viewing the birth of agnosticism from a perspective informed by these factors tends to assume that agnostics are inherently antireligious due to their scientific spirit and antagonistic to Christianity on account of their rabid loathing for the Bible. I have purposely de-emphasized the role of such factors so that we may view the origins of agnosticism through the development of their essential notion of the limits of knowledge. Huxley, Stephen, Spencer, Clifford, and Tyndall developed their agnosticism during the time when controversy raged over the issue of God's knowability, and their religious thought was shaped by those theologians who defended a modified Kantian position.

It was a notorious fact during the mid nineteenth century that German and English modes of thought could hold no commerce. But Mansel thought he could find a way to make oil and water mix, and it was partially through his work in this area that the agnostics learned how to synthesize German and English intellectual currents.

Chapter Two

MANSEL AND THE KANTIAN TRADITION

> *It has been observed by a thoughtful writer of the present day [Alexander Campbell Fraser, 1819–1914], that "the theological struggle of this age, in all its more important phases, turns upon the philosophical problem of the limits of knowledge and the true theory of human ignorance." The present Lectures may be regarded as an attempt to obtain an answer to this problem, in one at least of its aspects, by shewing what limitations to the construction of a philosophical Theology necessarily exist in the constitution and laws of the human mind.*
>
> HENRY LONGUEVILLE MANSEL

Henry Longueville Mansel was once described by William Whewell as "by much the most zealous English Kantian whose writings I have seen."[1] Considering that Whewell was one of the few Englishmen thoroughly familiar with Kant's work, and since he had dared to apply German modes of thought foreign to his fellow countrymen in *The Philosophy of the Inductive Sciences* (1840), Whewell's qualifications for judging the extent of Mansel's Kantian proclivities would seem to be unquestionable. But despite Mansel's avowed aim to instill in English minds a respect for Kant, his ambivalent attitude toward a number of central concepts of the critical philosophy led him to present a distorted caricature of the Kantian position on science, religion, and their interrelationship. Mansel may indeed have been the closest equivalent to Kant which Victorian England could produce, but his development of a philosophical viewpoint dramatically opposite to the thrust of Kant's thought is an ironical confirmation of the difficulty Mansel faced when he attempted to make the philosopher of Königsberg accessible to the English public.

The Life of a Controversialist

Mansel was born on 6 October 1820 at the Northamptonshire village of Cosgrove, where his father served as rector. The majority of Mansel's ancestors were important soldiers and clergymen, and the Mansels could lay claim to ancient and honorable descent. Burgon reports that the family traced their roots to a Philip le Mansel, who accompanied William the Conqueror into England in the eleventh century.[2] The young Henry was proud of his heritage, displayed an ardent love for High Church principles and Toryism, and intended to follow in his father's footsteps by entering the ministry. After a distinguished undergraduate career at St. John's College, Oxford, Mansel took his bachelor's degree in 1843. He was ordained deacon in 1844 and ordained priest the following year. But Mansel chose to stay within the ivory towers of academe at Oxford, and from 1843 until 1855 he earned his living as a private tutor. He was appointed to a series of academic positions of increasing importance and prestige, Reader in Moral and Metaphysical Philosophy at Magdalen College in 1855, Waynflete Professor of Moral and Metaphysical Philosophy for Oxford in 1859, and finally Regius Professor of Ecclesiastical History in 1866.[3] But his duties at Oxford weighed heavily upon him, and he was distressed at the liberal direction the university seemed to take, so two years later Mansel accepted Disraeli's offer of the deanery of St. Paul's. Mansel's life was cut short, for he died suddenly in his sleep on 30 July 1871 at the age of fifty-one, from a ruptured blood vessel in the brain.

Considering Mansel's distinguished reputation at the height of his popularity, his social and intellectual activity has received scant attention from historians.[4] During the 1850s Mansel was a respected figure at Oxford, regarded as one of the best teachers by the undergraduate students, and looked upon as a witty conversationalist by all who frequented the senior common rooms. He had established a name for himself as a logician of repute in the early fifties, and soon after, he demonstrated his skills in the realm of ethics and metaphysics.[5] He was asked to deliver the Bampton Lectures in 1858 precisely because Oxford High Churchmen believed that Mansel was potentially a new Butler who could offer a novel apologetic of intellectual substance to fill the void left by the Tractarians. An obsolete Pusey was no longer attractive to young minds, and conservatives were uneasy that the old man was the only counteracting force to the growing power of liberalism at the university.[6]

The situation at Oxford was only a microcosm of the predicament in which conservative Christians found themselves during the mid-Victorian period. Mansel from the start had seen that the larger

Henry Longueville Mansel
An Obituary Portrait

danger infecting the university community was coming from two sources. On the one hand, the Positivism of Comte's early thought seemed to Mansel to have crossed the channel to join hands with a native empiricist and scientific spirit that culminated in an implicit atheism. Mansel was particularly hostile toward John Stuart Mill and his *System of Logic* (1843) as a reflection of this intellectual development. On the other hand, Mansel perceived a second threat to Christianity in the slow permeation of German thought into England which brought with it both pantheism and biblical criticism. German pantheism denied one of Mansel's most cherished beliefs, the personhood of God, while the higher criticism offended Mansel's belief in the sacredness of the Holy Scriptures. In addition to attacking German thinkers such as Feuerbach, Fichte, Hegel, Schelling, Schleiermacher, and Strauss, Mansel criticized those Englishmen infected by the Germans: W. R.

Greg, Francis Newman and other unbelievers of the fifties, as well as the Broad Church.

Described as being "to the backbone a Conservative" by Burgon, Mansel's politics were a part of his religion.[7] For Mansel, Mill's empiricism, German pantheism, French Positivism, and biblical criticism were all the philosophical manifestations of the liberal movement in the political world. His response to the growing power of middle-class liberalism was similar to other orthodox reactions, such as Tractarianism and Ritualism, and was characterized by a mistrust of reason in the religious sphere and an emphasis on authority in the form of the Church or the Bible. But although Mansel shared the same aims as conservatives of this type, his Bampton Lectures were roundly attacked by men of his own party and judged by them to be an abysmal failure. From the time he delivered the Bampton Lectures up until his death thirteen years later, Mansel found himself almost constantly embroiled in controversy. In addition to answering to fellow conservatives, he was called upon to defend the position articulated in *The Limits of Religious Thought* from the attacks of a number of eminent liberal thinkers, including F. D. Maurice, Goldwin Smith, and John Stuart Mill. After his death his name sunk into obscurity, and his works placed in "a kind of 'Index Expurgatorius'" by Anglican thinkers, even though Mansel's orthodoxy was defended by his former students.[8] This was a tragic fate for one who had labored so hard during his lifetime for the conservative Christian cause.

Mansel, Hamilton, and Kant

What the critics disliked most about Mansel's work was the foreign flavor, despite his avowed intention to use German modes of thought as a means to achieve orthodox ends. German thought in any form was looked upon with suspicion by English minds. In William Hale White's *Autobiography of Mark Rutherford,* the protagonist recalled that at the Dissenting College where he was to be prepared for the ministry, "the word 'German' was a term of reproach signifying something very awful, although nobody knew exactly what it was."[9] The distrust of all things German was a powerful sentiment in England despite a brief flurry of interest in German literature in the 1820s (exploited by Carlyle) and the impact of German thought on men such as Thomas Arnold, Hare, Coleridge, and Whewell.

Mansel was one of the few English intellectuals who, during the fifties, could read untranslated German theology and philosophy and understand it tolerably well. He was aware of the differences between the intellectual traditions of Germany and England that made it next to

Sir William Hamilton
A Photograph in the National Galleries of Scotland

impossible for German philosophy to be easily accessible to English readers. The English he characterized as inductive and empirical, while the Germans were transcendentalists and idealists. "What concord or fellowship can be hoped for," Mansel asked, "between the laborious induction which traces all ideas to sensation and reflection, and the 'high priori' method which deduces a theory of the universe from the innocent assumption that A is A, on the bold paradox that A is equally not A?"[10] Mansel saw his role as introducing developments in German thought to the English public so that they could better deal with the challenge it represented to the orthodox Christian faith. What complicates Mansel's relationship to Germany was his use of a particular strand of German philosophy to undermine the claims of thinkers hostile to orthodoxy. In fighting fire with fire, Mansel's critics claimed, he had burnt to the ground the very institution he was attempting to protect.

Mansel never concealed his debt to German ideas, and in the *Prolegomena Logica* he singled out two philosophers as instrumental in the formulation of his thought (x–xi). The first was the prominent Scottish philosopher Sir William Hamilton, who occupied the chair of logic at Edinburgh University from 1836 to his death in 1859. Hamilton's learning was considered by his contemporaries to be extraordinarily vast, and his erudite essays on French and German philosophy suddenly restored Scotland's intellectual reputation on the continent.[11] During the controversies surrounding Scottish university reform Hamilton played a significant role in the fight to retain the old, national heritage of a general and philosophical basis to education as against those who desired to anglicize schools of higher learning through a stress on specialization and the classics.[12] Hamilton exercised tremendous influence over his students at Edinburgh, even after he was stricken by paralysis in 1844. "The massive brow and the calmly observant eye were clouded," one former student recalled, "the articulation was defective and laborious; but he struggled bravely on; and the moral effect on the students of that shattered body sustained by an indomitable will was immense."[13]

One of the major aims of Hamilton's thought was to synthesize Kantian criticism with the philosophy of the Scottish school of common sense. Hamilton was bent on using the critical philosophy as a weapon against the pretensions of Fichte, Schelling, Hegel, and Coleridge to philosophical hegemony. Mansel was considered by his contemporaries to be Hamilton's foremost disciple. He made the Scottish philosopher's name one to be reckoned with in England during the fifties and sixties, not only by referring continuously to Hamilton in his published works, but also by co-editing, with John Veitch, a four-volume edition of Hamilton's *Lectures on Metaphysics and Logic* (1861–66).

The second thinker who, according to Mansel, played a crucial role in his intellectual development was Immanuel Kant, the great German philosopher of the late eighteenth century and professor of logic and metaphysics at Königsberg from 1770 until his death in 1804. Kant's life seemed so regular and uneventful that the inhabitants of Königsberg could set their clocks to his movements. But the outward monotony of Kant's life was deceptive, because with the publication of *The Critique of Pure Reason* in 1781 he began to produce a series of books which would revolutionize European thought. Mansel believed that Kant (and Hamilton) had taught him the importance of epistemology as a preliminary to all investigation. From Kant he learned that

Immanuel Kant
A Photograph by Gottheil von Sohn of a Portrait by Doebler, 1791

"the true lesson of philosophy" is "a knowledge of the limits of human reason"(*LPK*, 4).

Mansel made Hamilton's ideas and his interpretation of Kant part of the intellectual scene of the fifties and sixties. In the process he helped to bring a new Kant to England. According to Mansel, Kant had never received his due in his own country, partly because the German thinker was heir to the English philosophical tradition as founded by Locke and Hume. Mansel asserted that Kant was "the philosophical offspring of Locke and Hume; his writings are the natural supplement and corrective of theirs; and it may be that the spirit of philosophy is not so extinct among the countrymen of Locke and Hume, but that the 'unsightly root' of the German sage may yet bear in another soil the bright golden flower which it has failed to produce in its own."[14]

Mansel's emphasis on Kant's affinities with the British empiricist tradition contradicted the prevailing English view of Kant. Most English thinkers of the first half of the nineteenth century perceived Kant as a transcendental philosopher. This is true both of the English romantics, like Coleridge, who were first to react positively to Kant's philosophy, and of the British empiricists, whose hostility toward Kant stemmed from their uncritical acceptance of the romantic view.[15] But it would be fair to say that Kant generally had little influence on English thought until the middle of the nineteenth century.[16] Only in the sixties did Kant's theories begin to penetrate English Christianity, which had been able to isolate itself from the influence of German and French thought during the first half of the century.[17] It was therefore necessary for Mansel to point out in 1853 that "it would probably astonish some of the critics who talk so comprehensively of German Metaphysics and German Theology, as if all Germans held the same opinions, to be told that the purport of the philosophy of Kant is to teach a lesson of humility, to inculcate the very limited nature of human faculties and human knowledge."[18]

Although Mansel admitted that he owed much to Kant (in particular, to Kant's *Critique of Pure Reason*), he was highly critical of many aspects of the German philosopher's thought. Mansel believed that some sections of Kant's works, especially those concerning "the right use and legitimate boundaries of reason," were brilliant, while others tended toward vagueness and error. Kant's obscure language, according to Mansel, often led thinkers to overlook or misunderstand the spirit of the whole, which taught that there are limits to human knowledge. For this reason Mansel often looked to Hamilton's work for a clear and consistent modification of Kant's theory of knowledge.[19]

Kant and the Enlightenment

Mansel hoped that by using the ideas of both Kant and Hamilton, in particular their epistemological concepts, he could simultaneously undermine German pantheism and French and English empiricism. However, in the realm of philosophy of nature Mansel's application of Kantian and Hamiltonian epistemology led him to a fundamentally sceptical stance that destroyed the possibility of certainty in science. Kant's main aim was the construction of an epistemology that would account for the possibility of certainty in natural science and yet could retain a place for religion and ethics. He was responding to the threat posed by Enlightenment scepticism and determinism.

By the middle of the eighteenth century the attempts of scientists and philosophers to discover the purpose of nature and the rights and duties of human beings solely through the use of human reason seemed to have led to a choice between the inscrutable and the intolerable.[20] The work of Newton and Locke, which had at first seemed to promise human liberation through a mastery of nature, was developed in such a way by Enlightenment figures as to present late-eighteenth-century intellectuals with two dead ends. They could either follow Helvétius and d'Holbach and preserve science through a conception that the universe is regulated by the blind determinism of matter in regular but aimless motion, or they could side with Hume and save ethics by adopting a sceptical position that denies that people can have access to objective knowledge. But the first option was fatal for religion and the second alternative destroyed the certainty of science. Kant wanted to find a way out of the Enlightenment impasse. Most of all, he desired to construct a third option that would reconcile and unify science and religion. Kant believed that he could accomplish his task if he merely unraveled the implications arising out of the philosophy of nature embedded in the original Newtonian view.

As Kant constructed a philosophy of nature upon which to erect a sound philosophy of science, the example he constantly kept before his eyes was the work of David Hume. In *A Treatise of Human Nature* (1739), in his conclusion to book one, "Of the Understanding," Hume admitted that he had not succeeded in uncovering a valid criterion of truth in his analysis of the understanding, but nevertheless he elected to proceed onto his examination of the passions and morals. Fretting over the corner he had painted himself into, a choice "betwixt a false reason and none at all," Hume regained his joviality when he recalled the solution provided by Nature. "Most fortunately it happens," Hume declared, "that since reason is incapable of dispelling these clouds, Nature herself suffices to that purpose, and cures me of this philosophical melancholy and delirium, either by relaxing this bent of mind, or by some avocation, and lively impression of my senses, which obliterate all these chimeras. I dine, I play a game of backgammon, I converse, and am merry with my friends; and when, after three or four hours' amusement, I would return to these speculations, they appear so cold, and strained, and ridiculous, that I cannot find in my heart to enter into them any further."[21]

Immersion in the inanities of social life, however, did not really provide Hume with a rational basis for philosophy. Passing time frivolously was no cure for the perplexities raised by his utterly ruthless consistency. Although Hume tried to mitigate his scepticism, his work amounted to an exposé of the weakness of Enlightenment sensationalism. Hume demonstrated that if one started from Locke's basic premise that knowledge is produced by experience received through sensations, then the legitimate conclusion to be drawn is the unreliability of such knowledge. Many of the axiomatic principles of Newtonian science, for example the existence of an external natural world, could not be philosophically justified by sensationalism. Experience was insufficient to determine whether the perceptions of the senses are produced by external objects resembling them. To Hume, Lockean empiricism was consistent only with scepticism.

It was in Germany that Hume's thought had the most profound impact. French and English thinkers never really confronted the problems raised by Hume and therefore did not realize that traditional philosophy had reached an impasse.[22] Kant recognized what Hume had accomplished, and he believed that the only response was the formulation of a revolutionary new program that started from non-Lockean assumptions. Instead of a philosophy of nature based on sensationalism, Kant developed a position in *The Critique of Pure Reason* anchored in what he called transcendental idealism.

Kant, Nature, and Transcendental Idealism

The Critique of Pure Reason is a search for knowledge "independent of experience and even of all impressions of the senses." In other words, Kant wants to restrict his study to a priori knowledge (42). Necessity and universality are the sure criteria of a priori knowledge, since experience can never give birth to this type of knowledge. Kant's guiding question throughout *The Critique of Pure Reason* is this: "Given *a priori* synthetic judgments, how are they possible?" Thus, he introduces a particular type of a priori knowledge by distinguishing between analytic and synthetic judgments. In analytic judgments, the connection of the predicate with the subject is thought through identity, while synthetic judgments are those in which this connection is thought without logical identity (48). Judgments of experience are synthetic and extend our knowledge. But how are a priori synthetic judgments possible if they attempt to extend our knowledge without the help of experience? Kant does not attack the problem using this question, rather he says *given* that we have a priori synthetic judgments, how are they possible? Science, for Kant, can only be built upon the secure basis of synthethic a priori knowledge, which is necessary and not merely contingent or empirical, and yet which is based on experience and not merely abstract. He points out that we are already in possession of this type of knowledge in the form of a priori synthetic judgments contained in the axioms of mathematics and natural science. The problem for pure reason to examine is the implications of our possession of this species of knowledge for the sensibility, the understanding, and reason.

Kant's sole consideration in the "Transcendental Aesthetic" is a transcendental doctrine of sensibility which will explain how pure mathematics is possible. His discussion centers around the question, What a priori representations does sensibility contain which constitute the condition under which objects are given to us? Kant is looking for pure forms of sensibility or pure intuitions "which, even without any actual object of the senses or of sensation, exists in the mind *a priori* as a mere form of sensibility" (66). He finds two, time and space. However, Kant is conscious that in making time and space "subjective" forms of sensibility, he has left himself open to the charge of idealism. He attempts to overcome this problem by maintaining both the ideality and reality of space and time. He accomplishes this by redefining the concept of objective reality as that which is universal to all people, and by rejecting the old conception of an independent, self-existing entity (things knowable as objects in themselves). Space and time are not things in themselves existing in nature independently of human beings, yet they are still objectively real (for people). In the case of time, Kant argues that "time is therefore a purely subjective condition of our (human) intuition (which is always sensible, that is, so far as we are affected by objects), and in itself, apart from the subject, is nothing. Nevertheless, in respect of all appearances, and therefore of all the things which can enter into our experience, it is necessarily objective" (77–78).

Now Kant is not talking of human beings as isolated individuals who organize the manifold of intuition into their own private idealistic illusions. Rather, Kant is discussing how people universally organize the matter of sensibility in order to communicate about nature. Therefore, when Kant uses the term "subjective," it is not meant to be the subjectivity of solipsism. The true aim is to break down the polar opposition between the terms "subjective" and "objective" or inner and outer. Just as "subjective" is not the illusion of solipsism, "objective" is not the absoluteness of the thing-in-itself. Nothing intuited in space or time is a thing-in-itself, for the a priori forms of sensibility do not inhere in things in themselves, rather they are the means by which our sensibility organizes the manifold of intuition.

The full implications of this view of time and space for a concept of nature and the status of the understanding were revealed in the "Transcendental Analytic." Nature is known only as appearance, and not as it is in itself or that which is completely unrelated to humanity. This notion of unrelatedness is meant to signify a conception of nature as that which exists independently of us or that which would subsist even if humanity, as a race, perished. "By *transcendental idealism,*" Kant explained, "I mean the doctrine that appearances are to be regarded as being, one and all, representations only, not things in themselves, and that time and space are therefore only sensible forms of our intuition, not determinations given as existing by themselves, nor conditions of objects viewed as things in themselves" (345). While on a transcendental level, man and nature (appearances) are somehow connected, empirically human beings represent objects in space as being external to them.

Kant maintained that the notion of transcendental externality excluded connection or relatedness but that holding to the concept of empirical externality allowed one to assert simultaneously that on the transcendental plane there is an interrelationship between human beings and nature. Transcendental idealists could admit that, on an empirical level, a dualism (self and not-self) existed without alienating themselves from nature on the transcendental plane. This dualism signifies that *from the empirical perspective* both self and not-self are real phenomenal entities and are really separate from one another in appearances (nature). Transcendental idealists are also empirical realists (or dualists), because they admit the existence of matter without having to resort to going outside their self-consciousness:

> For [the transcendental idealist] considers this matter and even its inner possibility to be appearance merely; and appearance, if separated from our sensibility, is nothing. Matter is within him, therefore, only a species of representations (intuition), which are called external, not as standing in relation to objects *in themselves external,* but because they relate perceptions to the space in which all things are external to one another, while yet the space itself is in us. (346)

Matter is perceived immediately as being external to us or in space on the empirical plane, but since space, from the transcendental perspective, is a form of intuition, we are related to nature (and nature is related to us), and thus we have our assurance of the empirical reality of nature. Kant stated that "external things exist as well as I myself, and both, indeed, upon the immediate witness of my self-consciousness" (346–47).

Kant believed that in maintaining this position, he could overcome Humean scepticism by avoiding the pitfalls of attempting to infer from an effect to a determinate cause:

> Now the inference from a given effect to a determinate cause is always uncertain, since the effect may be due to more than one cause. Accordingly, as regards the relation of the perception to its cause, it always remains doubtful whether the cause be internal or external; whether, that is to say, all the so-called outer perceptions are not a mere play of our inner sense, or whether they stand in relation to actual external objects as their cause. At all events, the existence of the latter is only inferred, and is open to all the dangers of inference, whereas the object of inner sense (I myself with all representations) is immediately perceived, and its existence does not allow of being doubted. (345)

In viewing appearances as an effect, one is unable to reach the cause. If inner sense causes empirically external things, then everything is a Berkeleian illusion ("a mere play of inner sense"). Each individual would create his or her own dreamworld. Yet also, if transcendentally external objects cause appearances, then we are unable to get to the transcendental cause, which is outside us (i.e., it is an unrelated thing-in-itself).

But Kant felt he had overcome this difficulty. "In order to arrive at the reality of outer objects," Kant claimed, "I have just as little need to resort to inference as I have in regard to the reality of the object of my inner sense, that is, in regard to the reality of my thoughts. For in both cases alike the objects are nothing but representations, the immediate perception (consciousness) of which is at the same time a sufficient proof of their reality" (347).

The epistemological position that ultimately had to resort to inference (from cause to effect) was referred to by Kant as transcendental realism/empirical idealism. In opposition to transcendental idealism, the transcendental realist regards "time and space as something given in themselves, independently of our sensibility" (346). Similarly, nature (outer appearances) is interpreted as a thing-in-itself, which is a self-existing, independent entity unrelated to us or our sensibility. By positing the existence (reality) of two entities on the transcendental level, the transcendental dualist (or realist) ends up dealing with the relationship between human beings and nature in terms of cause and effect.

The transcendental realist afterward plays the part of empirical idealist. "After wrongly supposing that objects of the senses, if they are to be external, must have an existence by themselves, and independently of the senses," Kant argued, "he finds that, judged from this point of view, all our sensuous representations are inadequate to establish their reality" (346). In making nature a thing knowable in itself, the transcendental realist/empirical idealist is unable, like Hume, to be certain of the reality of external objects of the senses. The end result is that the transcendental realist/empirical idealist is forced to resort to inference which never attains its object.

Kant's transcendental idealism not only was a position from which the axiom of an external world of nature could be justified, it also served to resolve Humean difficulties with other a priori synthetic judgments of reason such as the universal and necessary concept of cause. Kant maintained that only if nature is conceived of as being related in some way to human beings (i.e., only if nature is viewed as appearances) could the validity of the foundational propositions of science be made explicable. "Should nature signify the existence of things in themselves," Kant stresses, "we could never know it either *a priori* or *a posteriori*."[23] Kant argued that if nature is the thing-in-itself, as the transcendental realist contends, we could never have knowledge of it. On the one hand, the possibility of a priori knowledge of nature (conceived of as the thing-in-itself) is ruled out, for our understanding would have to conform to objects. On the other hand, a posteriori knowledge of nature (thought of as an independent, self-existing entity) is implausible because we could never know the laws of nature from experience, since necessity is derived solely a priori.

The fact, therefore, that we have scientific knowledge can be explained only if the human mind participates in constructing appearances. One can justify synthetic a priori judgments, upon which science depends, only from the transcendental idealist/empirical realist perspective. Concepts such as cause and substance are categories through which the understanding organizes the vast amount of sense data that is constantly fed into our minds. The synthetic a priori categories of the understanding allow us to "make sense" of the world in which we live by yielding knowledge of objects of the empirical world. The mind, in using the categories, is actively engaged in making experience possible through its systematic ordering of the manifold of sense data. Kant rejects the sensationalist assumption that the mind is a passive receptor of sense data because it undermines the very certainty and universality of science celebrated by the Enlightenment.

The Hamiltonian Natural Realist as Kantian Transcendental Realist

Mansel's mentor, Hamilton, was notorious for his aversion to physics and mathematics. Hamilton believed that the study of science could lead to materialism and atheism, and he was not concerned if his "philosophy of the conditioned" did not provide sound justification for the axioms grounding natural science. Although Mansel did not openly voice his hostility toward science, he, too, was uninterested in the implications of his position for science. In fact, Mansel fell into the contradiction of transcendental realism/empirical idealism because of his rejection of Kant's conception of nature as appearance. Mansel attacked Kant's transcendental idealism in two ways: first, by maintaining that it is impossible to know whether nature is appearance or not; and second, by espousing Hamilton's doctrine of natural realism.

Mansel asserted that an answer could not be given to the question, "Do things as they are resemble things as they are conceived by us?"[24] Only one who is able to compare the two can provide a legitimate response:

> When Kant (Kritik der r.V. p. 49) declares that the objects of our intuition are not in themselves as they appear to us, he falls into the opposite extreme to that which he is combating: the Critic becomes a Dogmatist in negation. To warrant this conclusion, we must previously have compared things as they are with things as they seem; a comparison which is, *ex hypothesi*, impossible. We can only say, that we have no means of determining whether they agree or not. (*PL*, 74)

Mansel argued that if things in themselves are absolutely unknown, then we are unable to say whether they are like or unlike nature.

Mansel seemed, however, to display an insight into the unknowable when he supported Hamilton's epistemological theory of natural realism. Hamilton looked to Thomas Reid (1710–1796), and not Kant, for his conception of perception. Reid, a Scottish philosopher, is generally regarded as the founder of the Scottish school of common sense, of which Hamilton was a member. Hamilton argued that both Reid and Kant were reacting to Hume's scepticism, but that Reid was a better philosopher, especially with regard to a doctrine of perception. Hamilton valued Reid for positing the possibility of an immediate knowledge of material objects, and he criticized Kant's retention of the idealism of previous philosophers. In opposition to what he called Kant's "Cosmothetic Idealism," Hamilton held to natural realism, which posited the unconditional veracity of consciousness in testifying to the reality of both mind and matter. Hamilton believed that we have an immediate knowledge of the external world, and at times he stated that we intuit the thing-in-itself.[25] Hamilton's natural realism was a variation of what Kant called transcendental realism.

Mansel agreed with Hamilton that the consequences of idealism "can only be avoided by abandoning the Idealistic theory, and substituting a Natural Realism." Nature is a self-existent, independent entity, and true knowledge yields insight into the thing-in-itself. "The objects which I am capable of knowing exist whether I know them or not," Mansel declared, "and my knowledge is real only in so far as it corresponds to the actual constitution of the thing known."[26] By embracing Hamilton's natural realism, Mansel repudiated Kant's transcendental idealism.

It might seem contradictory to attribute to Hamilton and Mansel, two avowed champions of the principle of the "relativity" of knowledge, the view of nature as the thing-in-itself. However, the contradiction exists in their thought. Hamilton was led into this problem by his attempt to synthesize two antagonistic epistemological systems, those of Kant and Reid. Mansel struggled in vain to reconcile Hamilton's natural realism to the Kantian idea that natural objects are known only as they stand in relation to human faculties. In *Metaphysics* Mansel ties himself into knots on this complex issue. We cannot, he begins, be conscious of objects out of relation to our own cognitive powers. We are not sure that, if our faculties were altered, the same things would appear to us in the same form as they do now. "But, on the other hand," Mansel insists, "we have no right to dogmatize on the negative side, and to assume, with equal absence of ground, that things *are not* in themselves as they appear to us" (54).

Mansel was suspicious of Kant's idea of appearances and of his stress on the active nature of the human mind, for he believed that transcendental idealism led inevitably to the attempt of German idealists and pantheists to overcome dualism. With the example of the history of post-Kantian German thought before his eyes, Mansel defended the Englishman's trust in common sense and its testimony to the existence of subject and object.

Mansel's view of Kant as the father of German idealism and pantheism raises interesting questions about the Kantian tradition. The fact that two disparate movements of thought, English agnosticism and German idealism, claimed Kant for their progenitor is an indication of the extremely delicate nature of Kant's philosophy. If too much emphasis is placed on one aspect of his philosophy, for example, the active power of the human mind, and if Kant's empirical realism is ignored, the result is idealism as presented by Fichte. Kant guarded against this by making the use of the categories of the understanding immanent. This signified that speculative reason could make no synthetic a priori judgments when dealing with the level of the transcendental world and that pure reason must accept the empirical realism of objects in appearances. Mansel, however, overreacted against the German idealists' misinterpretation of Kant and consequently put too much stress on the empirical quality of Kant's *Critique of Pure Reason.* Mansel denied that speculative reason was capable of making synthetic a priori judgments in order to guard against pantheism and idealism. But the result was the destruction of those necessary, a priori principles upon which Kant grounded natural science.

Mansel the Empirical Idealist

Mansel, in affirming the views of a transcendental realist, also held to the characteristic principles of empirical idealism. Mansel tried to avoid destructive idealism as best he could, but the logic of his position led him to admit that our senses do not give us information about an external real world.[27] Furthermore, Kant's a priori categories of the understanding, and the forms of the sensibility, were neither acceptable to Mansel, nor justifiable on Mansel's assumptions.

Several times throughout his works, Mansel attacks the association psychologists for denying the importance of a priori laws of thought. He asserts that one must account for the necessary truths of arithmetic and geometry and that this is impossible on the associationist theory. The necessity of mathematical judgments "results from the existence in the mind of the *a priori* forms of intuition—Space and Time." Mansel agrees with Kant that geometry and arithmetic contain a priori synthetic judgments that point to space and time as subjective conditions of all sensibility. Space is "a subjective condition of all sensible perception, and not a mere empirical generalization from a special class of phenomena," and this is evident from the fact "that it is impossible, by any effort of thought, to contemplate sensible objects, save under this condition."[28] However, in spite of these statements, which appear to be Kantian, Mansel's whole concept of space and time is very different from Kant's.

Mansel asserts that "we cannot help experiencing" a priori intuitions due to "our constitution and position in the world" (*PL,* 157). Kant would have argued that to explain the a priori forms of sensibility in terms of our "position" in the world mistakenly allows an empirical element to creep into the discussion. "Position," in the sense of location, is a contingent factor, and this type of consideration is not relevant to the search for necessary, a priori judgments. But more important, to maintain that space and time are a priori because "we cannot help experiencing" them is to imply that space and time are external, self-existing, and independent entities in the world with which we continually come into contact. Our unavoidably constant experience of these entities allows us to develop a corresponding permanent intuition that becomes an a priori form of sensibility. It is a priori, to Mansel, not because our minds actively organize the manifold of sense data, but because of the passive experiencing of a permanent condition of the external world. Space and time, to Mansel, do not make experience of the natural world possible, but experience of the natural world makes the a priori of space and time possible.

With respect to the categories and scientific knowledge, Mansel was even further from Kant's position. Pointing to flaws in Kant's logic, Mansel agreed with Hamilton's rejection of the categories of Kant's "Transcendental Analytic."[29] But Kant's categories, which were a priori synthetic judgments of the understanding, were meant as an expression of Kant's belief that nature is relational, or that, on the transcendental level, there is an interrelationship between man and nature. In dismissing the concept behind the categories, Mansel implicitly discarded the notion of appearances. Kant's transcendental deduction of the categories is not an empirical proof of their reality (for one is unable to turn to the empirical world for help in justifying synthetic a priori judgments). Rather, Kant's argument is constructed so as to reveal the consequences of each epistemological position. If empiricists do not want to admit a priori synthetic judgments, then the nature of knowledge becomes highly problematical for them. They are left with only purely analytic a priori judgments or a posteriori judgments upon which no certain science can be built. For there to be a certain science, human beings must participate in organizing the manifold of sensibility into appearances. In other words, people, through the categories, help make experience possible.

To probe deeper into Mansel's attitude toward Kant's categories and his justification of science, it is useful to examine how Mansel dealt with the specific category of cause and effect. Mansel affirmed that our concept of cause is derived from our consciousness of our own freedom. Our notion of the causal relation between two objects is modeled on the similar relation that exists between ourselves and our volitions. Mansel, however, questioned if the similarity is acceptable, and concluded that we cannot tell "how far the analogy extends, and how and where it fails." Cause, then, is not a necessary truth, nor is it "capable of any scientific application" (*PL*, 140–42). We delude ourselves into thinking that we know of the operation of cause and effect in nature because we illegitimately transfer a notion of cause acquired elsewhere. He concluded that we do not receive, anywhere, an intuition of cause in nature; therefore, since a concept of the understanding is valid only if it is based on a sensible intuition, cause is an illegitimate concept.

Mansel attacked Kant's concept of cause as a synthetic a priori judgment in the belief that he was preserving humanity's freedom and defeating Mill's deterministic stance in the *Logic*. Mansel was influenced in this strategy by Hamilton's similar ploy. Hamilton asserted that to accept a positive principle of causality was to accept fatalism, for we would find that everything, including human life, is caused (*LML* 2:412). But Kant had aimed to justify science while simultaneously preserving human freedom. He was able to achieve this through his distinction between appearances and the noumenal world. Cause and effect reigned supreme in nature, but people were free in the realm of noumena. This viewpoint was not open to Mansel, as he rejected Kant's notion of appearances. Once he had dismissed Kant's concept of different levels of existence, Mansel was left with a single existential plane where either necessity or freedom prevailed. Mansel had no other alternative, if he wished to be consistent, because to hold that freedom and necessity coexist within the same level of existence is to introduce a chaos that destroys both science (which requires the a priori necessity of cause) and ethics (which requires freedom). Mansel therefore had to choose between necessity and freedom because, from his position, only one would exist. Obviously, Mansel chose freedom, and the fact that he had destroyed science through his attack on the notion of necessity in cause and effect was of no consequence to him.

If we examine Mansel's attitude toward science, we will discover further confirmation of his empirical idealism. In rejecting the Kantian notion of nature as relational, he found no necessity in nature. He argued that "the belief in the uniformity of Nature is not a necessary truth, however constantly guaranteed by our actual experience."[30] Further, Mansel affirmed that "the fact that nature proceeds by uniform laws at all, is a truth altogether distinct from the laws of thought, and, if not of wholly empirical origin, at least one which cannot be ascertained *a priori* by the pure understanding" (*PL*, 208). In direct contrast to Mansel's views is Kant's belief that the categories make possible science, or the study of order in nature, through their active organization of the manifold of sensibility. "Thus the order and regularity in the appearances," Kant affirmed, "which we entitle *nature*, we ourselves introduce. We could never find them in appearances, had not we ourselves, or the nature of our mind, originally set them there" (*CPR*, 147).

Mansel may have learned of the importance of epistemology from Kant, but he appropriated only bits and pieces of the German thinker's theory of knowledge in his philosophy of nature. This sporadic borrowing from *The Critique of Pure Reason* gives Mansel's work a deceptive Kantian quality, and the illusion is only dispelled when their fundamental opposition is perceived. It may not seem significant that Mansel and Kant disagreed in their views on nature and science. However, Kant's religious thought was closely connected to the way he justified the validity of scientific axioms. His critical philosophy formed a systematic unity. In order to continue our evaluation of the accuracy of

Whewell's judgment on Mansel's Kantianism we will now turn to an examination of Mansel's religious thought. Mansel's divergence from Kant on science presented him with a variety of problems in his attempts to build a consistent philosophy of religion.

Kant, Practical Reason, and Religious Faith

Mansel claimed that the basis of his Bampton Lectures was the Kantian idea of the limits of knowledge. However, Mansel's motives for accepting the notion of the limits of knowledge differed profoundly from Kant's, and as a result, the thrust of their religious thought is diametrically opposed. Just as Mansel's philosophy of nature, despite its apparent Kantian flavor, was in the end the very transcendental realism against which Kant had warned as being fatal for natural science, in his philosophy of religion Mansel denied the validity of practical reason, which grounded Kant's religious and ethical position. As we follow Mansel in his rejection of the use of reason in religion and in his subsequent reliance on revelation, it will become clear that Mansel followed Hamilton in taking only the negative, destructive aspect of Kant's thought.

Having avoided the scepticism of Hume in his philosophy of nature, Kant did not desire to fall into the determinism of d'Holbach or Helvétius in his ethical and religious thought. In the "Transcendental Dialectic" of his *Critique of Pure Reason,* Kant addressed the issue of the connection between epistemology and religious ideas within the framework of a general discussion of how pure reason misuses the pure modes of knowledge of the understanding. Kant pointed to two situations where this takes place. The first arises if one tries to employ the categories of the understanding by themselves to comprehend appearances that are knowable only as the objects of possible experience (i.e., making a material use of the pure and merely formal principles of the understanding), and the second occurs when one attempts to apply the categories beyond the limits of possible experience. In both cases the sense data supplied by the intuition, upon which the understanding depends to produce real knowledge, is lacking. Pure reason falls victim to this error when it struggles to build a metaphysics based on a supposed speculative knowledge of God, freedom, and immortality.

In the specific case of theology Kant pointed out that "in all ages men have spoken of an *absolutely necessary* being, and in so doing have endeavoured, not so much to understand whether and how a thing of this kind allows even of being thought, but rather to prove its existence" (501). Kant argued that the very nature of the human mind and the way the mind constructs knowledge precluded the possibility of knowing God. He denied both that pure reason was capable of producing synthetic a priori knowledge of the transcendental realm, and that the categories were applicable to God. "Now as we have already proved," Kant asserted, "synthetic *a priori* knowledge is possible only in so far as it expresses the formal conditions of a possible experience; and all principles are therefore only of immanent validity, that is, they are applicable only to objects of empirical knowledge, to appearances. Thus all attempts to construct a theology through purely speculative reason, by means of a transcendental procedure, are without result" (529). The categories of the understanding were valid only if used immanently or phenomenally within the realm of nature (appearances), and they became sources of error when applied transcendentally or noumenally (i.e., to the thing-in-itself).

One purpose that lay behind Kant's insistence on the merely immanent employment of the categories of the understanding concerned his wish to protect God's status as a person. Kant perceived the importance of shielding all subjects, or persons, from the epistemological position that turns them into objects. Apply the categories to any entity, Kant would argue, and one is automatically relating to that entity as one relates to an object, no matter if the entity in question be subject or object. By attempting to know an entity through the categories of the understanding, one "objectifies" it. However, Kant, by restricting the use of the categories to appearances (i.e., by making the categories immanent), also limited knowledge to objects. The noumenal world is not to be thought of as an object of knowledge. God, as the ideal of pure reason and thus of noumenal quality, is transformed into an object of possible experience if we try to know him or prove his existence as an object of appearance (528).

In its search for totality and completeness, reason unknowingly falls into all sorts of contradictions. Kant warns continually of its inherent tendency toward falsehood, error, and illusion. But reason is also capable of guarding against deception. "The transcendental dialectic," Kant states, "will therefore content itself with exposing the illusion of transcendent judgments, and at the same time taking precautions that we be not deceived by it" (300). This is still a somewhat negative definition of reason's powers, but the purpose of *The Critique of Pure Reason* is largely negative and critical, because Kant's whole point in this work is to demonstrate the limited scope of pure or speculative reason. Kant viewed the first critique as clearing the ground for the second critique, which deals with practical reason, and it is in this realm that reason becomes constitutive by assuming a positive role. Practical reason could actively determine moral law or synthetic a priori judgments such as the ideas of freedom and the categorical imperative.

Kant's notion of practical reason is meant to be the supreme achievement of his thought. It is from the standpoint of practical reason that we are able to give meaning and significance to human existence by acting according to the categorical imperative. People are able to think of (not know) themselves as participating in the noumenal world by virtue of their freedom.[31] Kant conceives of the noumenal world (which he equates with the intelligible world and the realm of freedom and practical reason) as a platform from which we may interpret existence. Yet we cannot see behind the platform, nor does the platform enable us to transcend existence totally. We must never forget that we are a finite part of the physical world.

Kant's preservation of human freedom, and then his attempt to ground ethics in that freedom, are intimately connected to his religious thought. The moral principles arrived at by practical reason become, in his *Religion within the Limits of Reason Alone,* the criteria for evaluating the purity of the individual's concept of God, which in turn is the chief means by which we recognize an authentic revelation of the divine. For Kant theology is dependent on ethical theory. "Though it does indeed sound dangerous," Kant declared, "it is in no way reprehensible to say that every man *creates a God* for himself, nay, must make himself such a God according to moral concepts . . . in order to honour in Him *the One who created him*" (157).

When it came to the issue of revelation, Kant emphasized that people cannot accept the Bible blindly, but always bring to their reading of this text an interpretative guideline—an a priori idea of God. "Hence there can be no religion springing from revelation alone," Kant maintained, "i.e., without *first* positing that concept, in its purity, as a touchstone. Without this all reverence for God would be *idolatry.*" Kant held that the Bible was not automatically infallible and that it was to be judged by reason to determine if it accorded with the moral law. In contrast to a historical faith that requires scholars as intermediaries, Kant looks to "the pure faith of reason," which "stands in need of no such documentary authentication, but proves itself" (157, 120).

Dogmatic Scepticism and the Fideist Tradition

Kant's whole approach to theology was viewed with suspicion by the orthodox. Nevertheless, Mansel believed that elements of the critical philosophy could be pressed into service by Christian theologians with great success. The defense of conservative Christianity through the adoption of weapons drawn from thinkers perceived to be unorthodox has not been an uncommon occurrence in the history of Christian theology. When people flocked to St. Mary's in 1858 to hear Mansel deliver what Chadwick has called "the most instructive lectures of the century," they were usually treated to a Biblical quote at the beginning of each session.[32] Lecture three started off with the reading of Exodus 33:20–23, which Mansel offered as representative of a strand in Christian theology which always had been important. "And he said, Thou canst not see my face; for there shall no man see me, and live. And the Lord said, Behold, there is a place by me, and thou shalt stand upon a rock: and it shall come to pass, while my glory passeth by, that I will put thee in a clift of the rock, and will cover thee with my hand while I pass by: and I will take away mine hand, and thou shalt see my back parts; but my face shall not be seen" (*LRT*, 45). Even Moses, the most favored by God, leader of God's chosen people, was unable to see the divine visage. Mansel was not alone in interpreting this quote as an illustration of the limits of our ability to know a mysterious, transcendental deity.

I have already discussed the sceptical tradition and those hostile to traditional religion in the eighteenth and nineteenth centuries. In the beginning of this chapter I examined Kant's response to philosophical sceptics who doubted the validity of natural science. Mansel's Bampton Lectures point to a third type of scepticism that attempts to defend established Christianity by claiming that absolute knowledge is unattainable. This Christian scepticism, usually referred to as fideism, bases all certainty on faith and attempts to demonstrate that, independent of faith, sceptical doubts can be raised about any claims to knowledge or truth found through the rational faculty. Fideism covers a broad spectrum of views concerning the relation between faith and reason. A fideist of the extreme right would advocate a blind faith or unquestioning acceptance of some revealed truth while denying to reason any capacity to reach the truth. But a more moderate fideist would set faith above reason or take the position that once certain truths are taken on faith then reason has a valid role in clarifying our beliefs and adding to our knowledge.[33] In dealing with the correct attitude toward God, fideists might deny that knowledge of God is possible but nevertheless state that through faith we can affirm our belief in his existence.

Elements of fideism can be found in a number of major Christian thinkers and traditions from all ages, including the *theologia negativa* of Christian mysticism as typified by the writings of Pseudo-Dionysius, St. Augustine and the numerous members of the medieval Augustinian family, the followers of Tertullian in the middle ages, Occam and other nominalists, and Luther and Calvin. Popkin's work has concentrated on those Catholics who responded to the challenge of Protestantism by developing a form of fideism based on a marriage of Christianity and the sceptical tropes of Pyrrho. The French Counter-Reformers argued that the Reformers were dogmatically making reason the rule of faith, and then they formulated, with the help of Pyrrho and Montaigne, a scepticism with regard to the use of reason in religion. Popkin has also placed Bayle, a faithful Protestant, within the fideist tradition even though many of the Enlightenment philosophes found in Bayle's works valuable ammunition to be used against theologians and metaphysicians. It was not just sceptics of the sixteenth and seventeenth centuries who were sincere believers in Christianity, for Catholic apologists of the eighteenth century fought the rationalism of the philosophes by debunking the validity of reason and stressing the authority of faith. As Palmer argues "on the really fundamental question, whether man may trust his own mind to conduct him through the world, it was the religious believers in the eighteenth century, not the infidels, who were fatally touched with doubt."[34]

Writing on the measures taken by Christianity to defend itself during the nineteenth century, the rationalist Benn remarked that scepticism was the great bulwark of religious faith against the onslaught of rationalism. Indeed, Lamennais had adopted a form of fideism to answer the French Enlightenment and to defend conservatism and orthodoxy. But Benn was more concerned that "in England scepticism has become, under a modified form, the chief official weapon of official Christianity."[35] Besides Mansel, who is explicitly mentioned, perhaps Benn had in mind the Oxford noetics who, in the early nineteenth century, cast doubt on the ability of unaided human reason to attain truth in religious matters in order to gain greater freedom in the discussion of Christian dogma. Or, more likely, Benn might have been thinking of the Tractarians, who opposed the noetics by using the notion of the impotence of human reason in the theological realm as the justification for a stress on the authority of the Church.

Representative of the Tractarian viewpoint on this matter were the Anglican sermons of John Henry Newman (1801–1890), who, in 1845, traumatized the Oxford Movement by converting to Rome.[36] One theme that ran throughout Newman's work even after he had become a Catholic dealt with the danger of a reason that claims for itself full independence and the right to oppose faith. In his *Apologia Pro Vita Sua* (1864) he identified liberalism as "nothing else than that deep, plausible scepticism of which I spoke above, as being the development of human reason, as practically exercised by the natural man" (200). To counter liberalism, Newman was continually arguing that reason, without the aid of the rock of faith, could only produce uncertainties and doubt.

Twentieth-century Christian thinkers have found fideism an attractive position. Christian existentialists such as Marcel, and neo-orthodox theologians such as Barth, have stressed that God is "wholly other," and therefore that knowledge of God is impossible for finite human beings.[37]

The Ontological and Psychological Approaches to a Philosophy of Religion

Mansel's use of sceptical arguments to undermine the power of reason and simultaneously justify the dominion of faith was squarely within the fideist tradition of Christian theology. His distrust of rationalism and emphasis on the authority of the Church and the Bible was shared by conservative Christians of the nineteenth century such as John Henry Newman and the members of the Oxford Movement. But the scepticism that provided Mansel with ammunition was Kantian, and many English Christians were suspicious of the German philosopher from Königsberg.

For Mansel the virtues of the Kantian tradition lay in its emphasis on the importance of constructing a sound theory of knowledge as a starting point for all religious systems. "The primary and proper object of criticism," Mansel argued, "is not Religion, natural or revealed, but the human mind in its relation to Religion." In order to study the limits of religious thought, which was "an indispensable preliminary to all Religious philosophy," one first had to investigate the limits of thought in general (*LRT*, 16–17). To Mansel, the necessary laws of thought held for all subjects of thought, including religious topics. Here Mansel was adhering to Hamilton's assertion that "no difficulty emerges in theology, which had not previously emerged in philosophy," a dictum that Mansel affixed to the beginning to his *Limits of Religious Thought.* Hamilton rarely applied his epistemological theories to religious thought, although he hinted how fruitful this would be. Mansel endeavored to use Hamilton's theory of knowledge in dealing with problems in the philosophy of religion. In Mansel's opinion, Hamilton's "philosophy of the conditioned" should have been considered "the handmaid and the auxiliary of Christian Truth" (*LPK*, 44).

Mansel also felt indebted to Kant for teaching him to perceive the connection between epistemology and religious thought. He credited Kant with revolutionizing metaphysics by drawing attention to the importance of a psychological or epistemological approach in contrast to the old ontological viewpoint. "Instead of asking what are the circumstances in the constitution of things," Mansel argued, "by virtue of which they present such and such difficulties and contradictions to human understanding, we must ask what are the circumstances of the human understanding itself, by virtue of which a distinction exists between the conceivable and the inconceivable. Such, in fact, was the revolution introduced by Kant into metaphysical speculation" (*PL*, 75). Mansel found this distinction between the ontological and psychological method to be extremely helpful, and he used it as a basis for both the structure of his *Metaphysics* and the central argument of *The Limits of Religious Thought*.

In *The Limits of Religious Thought*, Mansel suggested that there are only two methods by which a rational religious philosophy may be attempted: the objective or ontological approach and the subjective or psychological method. Mansel conducted an examination of both methods in order to support the contention that reason could not construct a philosophy of religion on its own. In his analysis of the first, he claimed that metaphysicians and theologians based the objective or ontological method on a knowledge of the real nature of God. Mansel held that the ontological method described God as being the Absolute, the Infinite, and the First Cause. In the second chapter of *The Limits of Religious Thought*, Mansel dissected these terms as applied to God and concluded that, if attributed to one being, they become paradoxical. "But these three conceptions," Mansel declared, "the Cause, the Absolute, the Infinite, all equally indispensable, do they not imply contradiction to each other, when viewed in conjunction, as attributes of one and the same Being? A Cause cannot, as such, be absolute: the Absolute cannot, as such, be a cause" (31). The fundamental conceptions of rational theology are self-destructive. Even if taken singly, they lead to insoluble difficulties (33). Reason does not yield knowledge of God's nature, rather, it produces a startling array of contradictions.

This strategy of revealing the impotence of reason in the transcendental realm by illustrating its paradoxical results resembled Kant's section on the antinomies of pure reason in the "Transcendental Dialectic." Here Kant discussed how reason erroneously attempts to build a pure, rational cosmology by applying to appearances "that idea of absolute totality which holds only as a condition of things in themselves" (448). In so doing, one is able to "prove" two contradictory statements, such as "everything is composed of the simple/nothing is simple," or "the world is finite/the world is infinite." Kant's aim was to demonstrate that whenever reason produces paradoxical statements that seem equally reasonable, it has transgressed its own limits. However, Mansel was more concerned with internal contradictions in the concept of

God, while Kant was exploring contradictions arising out of the attempt to deal with nature as a totality. The common element is their stress on the illusory quality of the contradictions produced by reason.

Although it is clear that Mansel applied the basic idea of Kant's antinomies, it is important to note that here Kant was not dealing specifically with the concept of God. The substance of Kant's attack on rational theology was to be found in the discussion of the ideal of pure reason in the "Transcendental Dialectic," where Kant stressed protecting God from being turned into an object.

Mansel took only the negative side of Kant's "Transcendental Dialectic" and therefore emphasized the most destructive aspect of Kant's conception of the role of reason in religion. One must look to Hamilton to understand how Mansel modified Kant's thought. Hamilton took Kant's antinomies and created from them a law of the mind, referred to as the "Law of the Conditioned," which asserted that the conceivable was bounded by the inconceivable:

> the Conditioned is that which is alone conceivable or cogitable; the Unconditioned, that which is inconceivable or incogitable. The conditioned or the thinkable lies between two extremes or poles; and these extremes or poles are each of them unconditioned, each of them inconceivable, each of them exclusive or contradictory of the other. Of these two repugnant opposites, the one is that of Unconditioned or Absolute Limitation; the other that of Unconditional or Infinite Illimitation. (*LML* 2:373)

Hamilton claimed that since these two inconceivable contradictory extremes were "mutually repugnant, one or the other must be true" (*LML* 1:34). Mansel applied this "Law of the Conditioned" to his attack on rational theology by retaining Hamilton's purely philosophical terms "absolute" and "infinite" and giving them a theological significance.

Mansel argued that the ontological approach to a philosophy of religion was barren. His subsequent examination of the psychological method was to some extent an explanation as to why the ontological approach produced meaningless knowledge of God, for a study of the nature of the mind revealed the inevitable failure of reason to construct a valid religious philosophy. The ontological method, according to Mansel, tried in vain to reason from an object, God, down to human beings, but the opposing psychological method attempted to reason from the subject up toward God. While the former method was a branch of metaphysics known as rational theology, "the latter is a branch of Psychology, which, at its outset at least, contents itself with investigating the phenomena presented to it, leaving their relation to further realities to be determined at a later stage of the inquiry. Its primary concern is with the operations and laws of the human mind; and its special purpose is to ascertain the nature, the origin, and the limits of the religious element in man; postponing, till after that question has been decided, the further inquiry into the absolute nature of God" (*LRT*, 23). The psychological method, therefore, is epistemological, because it endeavors to find out what people can know about God given the nature of their minds.

Mansel's first step was to affirm that, given the structure of the mind, people can intuit God neither through the sensibility nor the understanding. In denying intellectual intuition Mansel was preventing the many liberal Anglicans influenced by Coleridge, who made reason into an intuitive faculty, from using this epistemological position as a point of departure for their religious thought. Mansel cleverly attacked the inspiration of Broad Church theology at its source by offering English intellectuals a new perspective on the Kantian tradition, which emphasized its empiricist leanings. But this position echoed the approach of Humean empiricism and did not explain fully the mass of confusion arising out of the ontological approach. Mansel, therefore, developed a critique of the conditions of consciousness or of what made thought possible. His main assumption was that the mind is compelled to think under certain laws that it cannot transgress. "If our whole thinking is subject to certain laws," Mansel stated, "it follows that we cannot think of any object, not even of Omnipotence itself, except as those laws compel us" (*PL*, 72). The aim of his argument was to prove that the key terms of rational theology abrogate the very laws of thought.

In particular, four conditions that make thought possible come into conflict with rational theology's conception of God. The first condition, that consciousness implies limitation or is only possible if we discriminate between one object and another, explains why we cannot understand the unlimited or infinite. Second, consciousness is only possible in the form of a relation between subject and object, and thus the absolute (that which is independent of all relation) cannot be thought without contradiction. Third, human thought is subject to time because consciousness involves succession in time, and hence the eternal (a timeless being) is inconceivable to us. Finally, the fourth condition of thought, that consciousness involves personality, which Mansel considered a limitation because we can only conceive of persons like ourselves, is further explanation of how rational theology's God is unthinkable.[38]

Mansel had learned a great deal about the limits of thought in general from Hamilton; this learning helped him develop his critique of reason from the point of view of the psychological method. Hamilton also saw the mind as bounded by laws of thought which set the unconditioned out of bounds (*DPL*, 14). Undoubtedly, both Hamilton and Mansel had appropriated those sections of *The Critique of Pure Reason* where Kant discussed how people cannot know God due to the nature of their minds. Mansel agrees with Kant that human beings can gain no knowledge of God through pure reason. Both argue that speculative reason is fallible in the realm of theology. But there are subtle differences in the content of the two arguments. Where Mansel bases his rejection of rational theology on laws of thought such as limitation and relation, Kant grounds his discussion on the inapplicability of the categories to God.

Reason and Faith

The profound differences between Mansel and Kant on religious thought are most apparent in the contrast in their attitudes toward reason. Mansel believed that his discussion of the ontological and psychological approaches to the philosophy of religion proved that the contradictions into which theology inevitably falls, when it attempts to conceive of God, exist in man's mind and not in God. Since it is not the nature of God but rather the nature of the human mind that is to blame, then one can still believe in God without knowing or comprehending him. By setting up a dichotomy between belief on one hand and conception, comprehension, knowledge, and thought on the other, Mansel tried to "prepare the way for a recognition of the separate provinces of Reason and Faith" (*LRT*, 39).

In adhering to this position, Mansel quite rightly saw himself as following Hamilton's lead. "The cardinal point, then, of Sir W. Hamilton's philosophy," Mansel claimed, "expressly announced as such by himself, is the absolute necessity, under any system of philosophy whatever, of acknowledging the existence of a sphere of belief beyond the limits of the sphere of thought."[39] Mansel's presupposition that God is totally transcendent or wholly other inevitably ended by giving human reason only a negative role in religion. Kant directly opposed the whole thrust of the standpoint that divorced reason from faith. Where Mansel grouped knowing, thinking, and reason together in opposition to faith, Kant conceived of thinking and reason as an integral part of the realm of faith.

Mansel was acutely conscious of his divergence from Kant on this matter, and throughout his works he attacked Kant as a rationalist. Those works in which Kant emphasized the importance of reason for the philosophy of religion, in particular *Religion within the Limits of*

Reason Alone, represented, to Mansel, Kant at his worst. Mansel stated that he would "rather contract than enlarge the limits assigned by Kant to the Understanding and the Reason." The English philosopher mistakenly saw Kant's insistence that human beings possess a faculty of reason as an attempt to build a new rational theology. Mansel asserted that after establishing an epistemology inimical to the ontological method, Kant proceeded to reconstruct what he had torn down. The nub of the problem, to Mansel, was that Kant viewed the understanding and reason as two separate faculties when in fact both were "governed by the same laws, and must be referred to the same faculty." Mansel argued it would be safer to conceive of reason as a mere impotence of the understanding.[40]

Mansel believed that by pointing to the defective nature of human reason he could undermine both pantheism and Positivism (including atheism, scepticism, and empiricism) in addition to rational theology, for he viewed them all as products of an exaggcrated use of reason. To Mansel, upholders of religion as diverse as Francis Newman, Kant, Fichte, Schelling, Maurice, and other Broad Churchmen relied on reason as much as did empiricists such as Comte and Mill. Mansel felt that his critique of a well-meaning rational theology could be extended into a condemnation of the use of reason altogether in defending or attacking religion. Adapting the strategy behind Kant's antinomies which had worked so well in his attack on rational theology, Mansel inspected pantheism and atheism. Mansel maintained that the theist used reason to prove that the infinite and the finite coexisted; the pantheist drew from the same source to deny the existence of the finite, while the atheist rationalized away the infinite. All three positions were equally irrational. "It is no matter from what point of view we commence our examination," Mansel claimed, "whether with the Theist, we admit the coexistence of the Infinite and the Finite, as distinct realities; or, with the Pantheist, deny the real existence of the Finite; or, with the Atheist, deny the real existence of the Infinite;—on each of these suppositions alike, our reason appears divided against itself, compelled to admit the truth of one hypothesis, and yet unable to overcome the apparent impossibilities of each" (*LRT*, 45).

Mansel admitted to the weakness in the theistic position but simultaneously pointed to analogous problems with atheism and pantheism. His whole aim in separating faith and reason was to permit himself to turn the weapon of atheism, empiricism, and scepticism (i.e., reason) back on itself. In admitting reason's impotence in religious matters (and subsequently relying on revelation), Mansel accommodated empiricism and scepticism within a religious framework. If empiricists argue that their five senses and their understanding do not intuit God, and their reason runs into difficulty proving God's existence, Mansel replies, "I grant you all of this." He would agree that a religious philosophy cannot be built solely upon these materials. But if sceptics try to use Mansel's attack on reason to advantage, they are frustrated in their attempt. True reason is deceptive in the religious realm, but this is due to our perverse misuse of our rational capabilities. "We may indeed believe, and ought to believe," Mansel states, "that the powers which our Creator has bestowed upon us are not given as the instruments of deception" (*PL*, 73). Reason is valid and competent if it does not stray beyond its limits.

A Positive Religious Philosophy

In granting the empiricists and sceptics their contention that reason is unreliable in religious matters and unable to prove God's existence, Mansel meant to move the emphasis away from reason as a means of constructing a religious philosophy and inexorably toward revelation. His negative and critical principles (from which the agnostics drew their epistemological theory) were a preliminary step to a positive and constructive attempt to build a philosophy of religion based on a psychological approach that did not use reason as its tool to know God's nature. In studying human beings, and not God, through his psychological method, Mansel endeavored to derive a notion of God founded on an investigation of human religious experience.

According to Mansel there are three sources from which we may form a judgment about the ways of God (*PC*, 145). The first source of information concerning God lies in a faculty of religious intuition upon which religious consciousness was built. The feeling of dependence and impotence leads us to a consciousness of God's power and impels us to prayer, while the sense of moral obligation establishes a belief in God's goodness and his role as a moral governor who is the source and author of the moral law within us. But Mansel maintained that these two elements of the religious consciousness were bound by the same limits as all consciousness and therefore did not reveal God's true nature. Mansel often complained that freethinkers such as Francis Newman depended too much on internal evidences. Similarly, Mansel limited the value of the second source, natural theology.[41] Natural theology was basically dependent on a reliable rational faculty that could justifiably draw an analogy from products of human contrivance to marks of design in the natural world produced by a divine intelligence. But Mansel had undermined both the power of reason and the idea that there was a close likeness between human and divine wisdom.[42]

Undoubtedly, Mansel saw the supreme achievement of his whole religious philosophy to be his defense of the third source of knowledge about God, revelation. Where Mansel seemed radical in his use of Kant and Hamilton to attack rational theology from the sceptical position, his true ultra-conservative colors show through when we see that the whole tendency of his thought is to uphold the dogma of biblical infallibility. Mansel sets up a dichotomy between revelation and reason that parallels the antithesis between faith and reason.

Mansel conceived of the Bible as a communication from a transcendent, infinite deity to a finite being. "Revelation," Mansel claimed, "represents the infinite God under finite symbols, in condescension to the finite capacity of man" (*LRT,* 20). Like the other two sources of information concerning God, revelation did not reveal God's true nature. The Bible presented humanity with regulative, not speculative, principles that "do not serve to satisfy the reason, but to guide the conduct: they do not tell us what things are in themselves, but how we must conduct ourselves in relation to them."[43]

From these premises, Mansel could argue that neither pantheists nor empiricists, nor any other type of thinker, could criticize the Bible by employing reason:

> If Revelation is a communication from an infinite to a finite intelligence, the conditions of a criticism of Revelation on philosophical grounds must be identical with those which are required for constructing a Philosophy of the Infinite. . . . Whatever impediments, therefore, exist to prevent the formulation of such a philosophy, the same impediments must likewise prevent the accomplishment of a complete criticism of Revelation. (*LRT,* 18)

One must possess a philosophy of the infinite in order to criticize the Bible, but since Mansel has ruled this out due to the existing laws of human thought, revelation is above criticism. A corollary of this position is Mansel's insistence that logically one must accept the whole of revelation. It was an all or nothing situation for Mansel. Either we accepted Christ as the Son of God, "and if so, we may not divide God's Revelation, and dare to put asunder what He has joined together,—or the civilized world for eighteen centuries has been deluded by a cunningly devised fable," and Christ was "an impostor, or an enthusiast, or a mystical figment; and his disciples crafty and designing, or well-meaning but deluded men" (*LRT,* 162). To Mansel, Christianity was wholly true or else wholly false.

Mansel's purpose was to undermine completely the basic assumptions behind biblical criticism and to attack the Broad Church and unbelievers of the forties and fifties for daring to think that they can improve revelation by simply chopping off those portions that displease them. Some Broad Churchmen and unbelievers such as Francis Newman and W. R. Greg had rejected the infallibility of the Bible on the grounds that God is portrayed as sanctioning immoral acts. Mansel objected that these men erroneously assumed that "the moral or intellectual nature of man is made the rule to determine what ought to be the revealed attributes of God, and in what manner they must be exercised" (*LRT*, 28). Later, Mansel was attacked vigorously on this point, for many felt that he was espousing a theory that denied that the term *goodness* possessed essentially the same meaning when applied to both people and God. This is analogous to maintaining that Mansel's regulative truth was speculative falsehood, which Mansel refused to admit. Mansel never really clarified the problem of how divine goodness resembles human goodness.

Interestingly enough, Mansel blamed Kant and his followers for the contemporary rationalistic tendency in England to criticize revelation on moral grounds:

> The works in which Kant and Fichte have attempted to construct an *a priori* criticism of revelation, upon moral grounds, are remarkable instances of this departure from the limits of all sound philosophy. Both assume that the sole purpose of revelation must be to teach them morality; and both assume that the morality thus taught must be identical to the minutest particular with the system attained by human philosophy; which last is supposed to be absolutely infallible. Hence Kant maintains that the revealed commands of God have no religious value, except in so far as they are approved by the moral reason of man.[44]

Mansel perceived in Kant's practical reason a morality that was obligatory for all rational beings, including God. This naturally put us and God on the same level as far as morality is concerned. Kant indeed maintained that "unless we wish to deny to the concept of morality all truth and all relation to a possible object, we cannot dispute that its law is of such widespread significance as to hold, not merely for men, but for all *rational beings as such*—not merely subject to contingent conditions and exceptions, but *with absolute necessity*." Thus, even God has to treat humankind as an end in itself. Mansel, however, would admit none of this, and he not only repudiated the idea of the categorical imperative but also charged that Kant's practical reason represented the attempt "to construct once more, in its most dogmatic form, that philosophy of the absolute which his criticism of the speculative reason was expressly instituted to overthrow."[45]

However, Mansel's rejection of practical reason, and the epistemological position upon which Kant preserved both science and religion, presented him with insurmountable difficulties in his attempt to frame a consistent ethical and religious philosophy. Mansel fell into the trap of conflating the realms of freedom and nature when he attempted to build a bridge from an ontological approach to psychology to a philosophy of freedom and a theory of divine morality. Although Mansel asserted that ontology is objective and psychology is subjective, he paradoxically maintained that psychology is a valid ontology because people are able to intuit themselves as the thing-in-itself. Mansel believed that he could move from psychology as a valid ontology toward a justification for freedom. Unless we are directly conscious of the self as noumenon, he argued, there can be no consciousness of having power over one's own determinations, and hence no freedom. In other cases where he deviated from Kant, Mansel was usually following Hamilton's lead. However, here this is not the case, as Hamilton asserted that mind is an unknown.[46]

It was important for Mansel to insist on knowledge of self and freedom, for he linked both to the existence of an intuitive moral faculty, implanted in human beings by God, which compelled them to "assume the existence of a moral Deity, and to regard the absolute standard of right and wrong as constituted by the nature of that Deity" (*LRT,* 74). Consciousness of God is possible, for Mansel, only if we are conscious of ourselves. Kant maintained that the "I" could not be intuited, for that would allow a human being to become a possible object of knowledge and therefore subject to the chain of cause and effect in appearances. Mansel's epistemology was unable to support the very notions, such as the importance of freedom, the distinction between persons and objects, and the existence of an intuitive moral faculty, that he tried to maintain in the face of attack from Utilitarians like Mill.

Agnosticism and Mansel's Kant

Mansel is best known for his attack on rational theology, and it is here that he seems closest to Kant and the agnostics. The agnostics were attracted to Mansel's insistence that reason be given a completely negative role in religious and transcendental matters. Nothing compelled them to follow out Mansel's second step, the reconstruction of a religious philosophy based on the religious consciousness, natural theology, and an infallible Bible. With the advantage of hindsight Flint recognized how Huxley and his friends later benefited from Mansel's use of Kant. This led him to the hasty conclusion that religious agnosticism, like "every kind of agnosticism[,] tends towards agnostic completeness." In other words, any use of sceptical principles in one sphere, be it religion, science, or philosophy, tended to demand a consistency that ultimately infected the whole body of thought with a debilitating scepticism. Flint referred to religious agnosticism as "inherently self-contradictory" and saw the alliance of agnosticism with fideism as "unnatural." He warned that the religious agnostic's denial of knowledge of God was far more dangerous than the antireligious agnostic's denial, since the latter is generally discounted and the former much overestimated. "The assaults of Sir William Hamilton, and Dean Mansel on the evidences or rational bases of theistic belief," Flint declared, "made a vastly greater impression on the public mind than those of J. S. Mill, W. K. Clifford, and G. J. Romanes."[47]

Flint's belief that scepticism is a very dangerous weapon that rebounds on its user holds for the Victorian agnostics and Mansel, but it would be a mistake to say the same for Kant. For Kant it is knowledge which is banished from the religious sphere. Reason gives rise to illusions but is not totally deceptive. Mansel distorted Kant's theory of knowledge by selecting only certain facets of the argument in *The Critique of Pure Reason* and by failing to understand how this book laid the groundwork for *The Critique of Practical Reason.* Instead of grounding his philosophical theology on practical reason's affirmation of God through human freedom and morality, Mansel espoused a belief in God based on the impotence, ignorance, and finite limitations of human beings. In his haste to protect revelation by valuing the authority of the Bible above human reason, Mansel took only the negative, destructive aspect of Kant's thought. It was Kant's insistence that we can have no knowledge of God through our understanding or speculative reason which Mansel adopted in his own religious philosophy. Mansel's selective use of Kant, and his additions, warped Kant's delicate epistemological viewpoint so that, in Mansel's hands, reason was divorced entirely from faith. Only through such a process could Kant's brand of agnosticism be transformed into the basis for what was to become Huxley's agnostic position.

In view of the opposition between Kant and Mansel on many of their key ideas, Whewell's claim that Mansel was the most "zealous English Kantian" stands as an ironic commentary on the inability of English intellectuals, including Mansel, to come to grips with the German mind. There were two Kants in mid-nineteenth-century England. On the one hand, there was the picture of Kant the transcendentalist glorified by the English romantics, Coleridgeans, Broad Churchmen and, later, the neo-Hegelians. But, on the other hand, there was the version of Kant as an empiricist who eschewed the use of reason in religion which was presented by Hamilton, Mansel, and Huxley. When the ideas of great thinkers are subjected to a post-mortem by those claiming to be philosophical descendants, more often than not the integral unity of their thought is dissolved in the process of analysis. Both pictures of Kant were distortions, the result of a one-sided emphasis on one aspect of his thought. English intellectuals of the nineteenth century seemed to be unable to hold together both Kant's criticism of pure reason and his construction of a theory of practical reason.

Similarly, English thinkers encountered difficulty when they attempted, as Kant had, to preserve science and religion. Actually, Mansel was not interested in justifying science; he saw Kant as a means to protect religion from the attacks of science. But he was not successful in providing a consistent theory of knowledge for this purpose, because his attack on rational theology seemed to the agnostics to undercut the very Christianity he defended. The agnostics therefore fastened onto Mansel's vision of the Kantian tradition in order to legitimize the value of natural science and their view of true religion. However, Kant's epistemology proved to be too difficult for them to control, and their modification of its sceptical element undermined natural science, while their rejection of practical reason weakened their attempt to allow a place for religion.

Chapter Three

HERBERT SPENCER AND THE WORSHIP OF THE UNKNOWABLE

> *. . . who could hold it against the agnostics if, as votaries of the unknown and mysterious as such, they now worship the* question mark itself *as God?*
>
> FRIEDRICH NIETZSCHE

The appearance of Mansel's Bampton Lectures in print in 1858 led to a controversy that lasted well into the sixties. Major participants included the Tractarian James B. Mozley (1813–1878), Broad Churchmen Frederick Denison Maurice (1805–1872) and Goldwin Smith (1823–1910), Scottish philosopher James McCosh (1811–1894), and unbelievers Herbert Spencer and John Stuart Mill.[1] The Mansel controversy was immensely important for the development of Victorian thought even though it is often overshadowed by the contemporary debates surrounding Darwin's *Origin of Species* and *Essays and Reviews.* Maurice referred to the publication of Mansel's book as "a critical event in the history of the English Church," while in 1859 the High Church journal *Literary Churchman* labeled the Mansel controversy "the great literary event of the year."[2]

The Mansel Controversy

The controversy involving Mansel's Bampton Lectures increased the interest of the public in the questions discussed and ultimately accustomed the English mind to thinking in terms drawn from German modes of thought. "The doctrine of religious nescience," James Martineau remarked in 1862, "has been rendered so familiar by Mr. Mansel, as to belong to the common stock of contemporary thought, and to make any full exposition of its grounds unnecessary."[3] It was now possible for the issue of the capacity of human beings to possess knowledge of God to become the subject of intense examination and discussion. By crystallizing complex epistemological arguments into this one question Mansel molded public debate among Christians and unbelievers for years to come. In 1884 one Christian apologist drew upon Mansel in order to investigate "the grounds on which God is said to be unknowable, and the grounds on which Christians assert that they may know Him. The question is one of much importance in these days: in fact, it may be said to be the question of the day."[4] Mansel was still setting the parameters of discussion as late as the eighties.

Mansel's manner of presenting the situation as a stark either/or led a number of important Christian thinkers to reconsider the virtues of the old fideist tradition. Although he was hailed as a champion of orthodoxy for a time, Mansel's position was slowly perceived as being far too extreme.

In an unsigned article in the *Rambler* in 1858, the liberal Catholic Richard Simpson insisted that in his Bampton Lectures Mansel had merely repeated the theme of John Henry Newman's *Parochial Sermons* without acknowledging his debt. Mansel apparently read this review, for he privately wrote Simpson and replied publicly in the preface to the third edition of *The Limits of Religious Thought.* Simpson wrote to Lord Acton on 6 April 1859 that "there has been an affectionate correspondence between Mansel and me; he has made a handsome speech about Newman in the 3rd edition of his Bampton Lectures." Mansel announced that he had never come into contact with Newman's book, but magnanimously admitted that a better acquaintance with Newman's works might have taught him "a better mode of expressing many arguments."[5]

Yet Newman, who had been following the Simpson-Mansel exchange with interest, felt uneasy that his earlier thought resembled this High Churchmen's controversial set of lectures. In December, 1859, he recorded his fears. "Mr. Mansell's [*sic*] doctrine has met with sufficient opposition among Protestant divines," Newman wrote, "to make me look narrowly to what I have myself before now said upon the subject which he treats." In this same letter, Newman went on to assert that, despite Mansel's claims, knowledge of God is "more than mere *relative* knowledge."[6] Newman denied that our knowledge of God through revelation is regulative; rather, he asserted that it is speculatively true as well as practically true. In turning away from the fideist strain of his earlier Protestant days Newman was moving closer to the position of the Roman Catholic Church in the nineteenth century. The capacity of human reason to prove with certainty central Christian beliefs, including the existence of God, was upheld by Pius IX in his encyclical *Qui pluribus* (1846), by Vatican Council I (1869–70), and by Leo XIII in the encyclical *Aeterni Patris* (1879).[7]

Mansel's fellow High Churchmen also drew back from the blunt conclusions of the Bampton Lectures. To Mozley they seemed "to put forward the absolute unintelligibility of the Divine nature—even Divine moral character—too nakedly." In his review of *The Limits of Religious Thought,* though anxious to find points of agreement between his position and Mansel's, Mozley insisted that human beings possess more than regulative knowledge of God because God has revealed himself "to His creatures, if at all, in a mode which is *speculatively* true."[8]

Liberal Anglicans such as Maurice, however, attacked Mansel most forcefully. The Mansel-Maurice controversy has been extensively examined, and Mansel's primary historical significance is generally located by scholars in his opposition to Maurice.[9] Since Maurice's faith was grounded on the belief that God constantly revealed himself in people's everyday lives, he perceived Mansel's denial of knowledge of God as a direct attack on the heart of his position. It was Maurice who pointed out the dangerous implications of Mansel's Bampton Lectures. "The confirmed, self-satisfied atheist is the one person who could receive such tidings without a protest, with perfect complacency," Maurice declared.[10]

The attacks on *The Limits of Religious Thought* bewildered Mansel. He was unable to comprehend how the idea of the unknowability of God, a central strand in past Christian theology, could be viewed as harmful to the cause of religion. In the fifth edition of *The Limits of Religious Thought,* published in 1867, he included a list of authorities, ancient and modern, whose testimony could be cited in support of his principal doctrines. Quoted were Clement of Alexandria, Origin, Cyprian, Arnobius, Athanasius, Cyril of Jerusalem, Basil, Gregory of Nyssen, Gregory Nazianzen, Chrysostom, Augustine, Aquinas, Hooker, Boyle, Stillingfleet, Leslie, Butler, Coleridge, and Bishop Browne, among others.[11]

During the last term of 1868 Mansel delivered a course of lectures on gnosticism in his capacity as Regius Professor of Ecclesiastical History. The gist of his argument centered on his claim that gnosticism was viewed by the early Christian Church as a heresy and that it was opposed by eminent theologians such as Irenaeus and Tertullian on the grounds that the genuine Christian believed in the "unsearchableness of God and the ignorance of man."[12] The lectures were published posthumously in 1875, and they represent Mansel's last attempt to convince his contemporaries, through a historical study, that his position in the Bampton Lectures was the true orthodox standpoint. Ironically, the year after the lectures were presented Huxley also pointed to the significance of gnosticism when he coined the term *agnosticism.*

The 1850s and 1860s saw the development of a new form of unbelief based on what was originally an important element in traditional Christian theology. The idea of God's unknowability became the fuel for a type of scepticism which was hostile toward Victorian Christianity, though not actually antireligious. Mansel unintentionally helped to found the Victorian agnostic school of thought by contributing to a resurgence of interest in epistemological topics in England.

Mansel's Unwanted Disciple

The appearance of Herbert Spencer's *First Principles* (1862) prompted Goldwin Smith to charge Mansel with encouraging the development of atheism:

> To prove that I am not guilty of calumny or the victim of hallucination in saying that the tendency of Mr. Mansel's doctrine is atheistical, I appeal to Mr. Herbert Spencer's work on "First Principles." Mr. Mansel will there find his own doctrines adopted in his own words as the foundation stone of a great system of philosophy, which he and I should agree in calling atheistical, by a very acute and honest writer.[13]

Mansel denied that Spencer's *First Principles* was a legitimate application of the Bampton Lectures. "I have not read Mr. Spencer's book on First Principles," Mansel asserted, "which I believe is only printed for his own subscribers; but from what I know of it indirectly, and from what I know directly of the author's other writings, I believe his teaching to be the contradictory, not complement, of mine."[14]

Despite Mansel's fervent desire to preserve a respectable distance from a thinker whose work he rightly considered as subversive of Christian orthodoxy, a number of contemporaries and scholars have viewed Spencer's *First Principles* as the logical next step to *The Limits of Religious Thought.*[15] Others, though agreeing with Mansel that Spencer's position was not the inevitable conclusion to be drawn from the premises of the Bampton Lectures, still view Mansel's *Limits* as the chief source of Spencer's agnosticism.[16] However, past treatments of the Mansel-Spencer connection, and Spencer's whole approach to epistemological questions, have been all too brief.

Today Spencer's work seems pompous, monotonous, and without interest, particularly since it was informed by scientific theories that are no longer considered valid. But recently, historians have emphasized Spencer's immense influence and importance in Victorian intellectual life as a corrective to the rather rapid decline his reputation suffered at the turn of the century.

There were a number of reasons for Spencer's popularity. His ten-volume *System of Synthetic Philosophy* (1860–96) marked him as the last in a long line of English polymaths like Herschel and Whewell. Spencer's magnum opus consisted of restatements, in systematic form, of recent theory and knowledge in all of the important areas of study, including volumes on biology, sociology, ethics, psychology, and religion. Furthermore, Spencer was endowed with great powers of synthesis, and he could gather the huge mass of information he was presenting into an integrated whole. Clearly, a new synthesis of all knowledge was attractive to those concerned with the fragmentation of the intellect into separate, specialized disciplines. The whole key to Spencer's ability to unify knowledge into a totally comprehensive philosophical system was his insistence that all phenomena be interpreted according to the law of evolution. This provided his work with an aura of scientific respectability. Although we distinguish between the evolutionary theories of Darwin and Spencer, the Victorian public routinely conflated their views. Spencer's Lamarckianism, ultimately in opposition to Darwin's stress on natural selection and his careful restriction of evolution to the biological realm, nevertheless gained scientific plausibility when it was linked with the *Origin of Species.*[17]

But above all, in addition to offering a synthesis blessed by science, Spencer built into the worldview that informed his whole system the social, political, and religious ideas that were familiar to the Victorians. Rarely has an intellectual read so little and produced so many volumes. The sources of Spencer's thought were the occasional book he did read (such as Mansel's Bampton Lectures), a fertile imagination, conversations with friends, and newspaper articles, all of which helped him to breathe the intellectual air of his times. Lauwerys has observed that "he acquired unconsciously a knowledge of what was being said and thought; and what he thus picked up, he gave back again in sonorous language and much amplified—swelled out, so to speak, to cosmic proportions."[18] The ideas he found "in the air" of mid-nineteenth-century England were those of the great Nonconformist middle class, who were transforming the country. Indeed, Wiltshire has argued that Spencer's scientific theory was merely the rationale for previously formed socio-political principles.[19] The scientific principle that the universe is evolving toward differentiation and individualization meant to Spencer not only that the principles of laissez-faire liberalism were embedded into the very laws of nature, but also that the harsh turmoil of rapidly developing industrial capitalism was an unavoidable side effect of an inevitable process. Spencer offered reassurance that this was the path to progress.

Spencer's system was also susceptible to a religious interpretation because of its affinities with certain Protestant theological creeds, and it proved an attractive alternative for those who repudiated Paley's brand of natural theology, the doctrines of depravity and perdition, and the belief in scriptural infallibility. Spencer retained in his universe an eternally transcendent and boundless power whose existence guaranteed the ultimate purpose and justice underlying the vale of tears of capitalist society.[20]

The Martyr of Science

The optimistic spirit of Spencer's philosophy is sadly at odds with the unhappy story of his life. Born in 1820 at Derby, Spencer was the only surviving child of William George Spencer, a schoolmaster of progressive educational views and a property owner. Herbert had a lonely and joyless childhood with no brothers or sisters and no schoolfellows until he was ten. Surrounded by the austere atmosphere of English middle-class dissent, with its emphasis on individualism and moralism, he grew introverted and intensely reflective.[21] Later, Spencer worked in a variety of jobs, first as a railway engineer (1837–46), then as subeditor of *The Economist* (1848–53), and finally as a freelance journalist in 1853. Overwork, financial insecurity, and a feeling of indecision took its toll in 1854 when Spencer suffered a nervous breakdown. For the rest of his life Spencer could only work in short spurts and frequently complained of both physical and mental illness. But in 1858, when he was correcting essays for republication, Spencer suddenly recognized the universality of their underlying assumptions. It was nothing short of a revelation for him. He decided to devote the remainder of his life to the systematic dissemination of his grand insight into the workings of the universe.[22]

Spencer totally absorbed himself in writing the *Synthetic Philosophy* for thirty-six years. All social intercourse was restricted to a bare minimum in order to avoid excessive excitement so disturbing to Spencer's concentration. Even his close friends had to work hard to convince him to join them for social occasions, and when he did indicate a willingness to attend he often only tentatively accepted an invitation. Spencer imagined himself to be of delicate constitution, and he kept a close watch over his condition on a daily basis.

Beatrice Webb, who became the prop of Spencer's declining years, and who was acknowledged by him to possess a unique insight into his character, made the following entry into her diary in 1884. "There is something pathetic in the isolation of his mind, a sort of spider-like

Herbert Spencer
A Photograph in the Spencer Papers at the University of London Library

existence; sitting alone in the centre of his theoretical web, catching facts, and weaving them again into theory. It is sorrowful when the individual is lost in the work."[23] Spencer's single-minded cultivation of the intellect in order to complete his task stunted the development of his emotional and aesthetic faculties, which in turn limited his insight into the human condition.[24] John Fiske, an American evolutionist who greatly admired Spencer, complained that Spencer "never seems to warm up to anything but *ideas.* He has got so infernally critical that not even the finest work of God—a perfect day—is quite fine enough for him. So he picked flaws with the grey-blue sky and the peculiar Turner-like light, and everything."[25] In 1896, after years of hoarding energy, of emotional deprivation, personal isolation, and sacrificing everything to his work, Spencer completed his *System of Synthetic Philosophy.* But by this time his achievements were already obsolete. Scientific knowledge had increased and specialized at a tremendous pace, leaving Spencer behind, and the England of middle-class dissent, which Spencer knew from the mid-Victorian period, and whose values were enshrined in his system, had been transformed into a world not to his liking.

"Well, we always have one consolation, such as it is," Spencer wrote to Huxley in 1888, when the latter was ill, "that we have made our lives of some service in the world, and that, in fact we are suffering from doing too much for our fellows" (ICST-HP 7:217). Spencer embraced his hermit-like existence and adhered to a life of asceticism not only for the sake of humanity but also in the name of a higher being. The laws of nature, and specifically the law of evolution, symbolized to Spencer the revelation of an immutable moral order. The theological element in Spencer's system can be understood in light of his Calvinist background, which left the indelible traces of Christian ideas on his thought. Like so many other agnostics, Spencer came from a family of earnest evangelicals. His parents were Wesleyan Methodists. However, in his *Autobiography,* Spencer asserts that he turned away from his parents' faith toward the end of the thirties when he was in his late teens. "My father's letters," Spencer recalled, "written during this period from time to time called my attention to religious questions and appealed to religious feelings—seeking for some response. So far as I can remember they met with none, simply from inability to say anything which would be satisfactory to him, without being insincere" (1:170).

There seem to have been a number of reasons for Spencer's early loss of faith. He reveals a feeling of resentment for the "foolish pertinacity with which, as a child, I was weekly surfeited with religious teachings and observances." This dislike of the oppressive atmosphere of his evangelical home led, he admitted, to a "certain disagreeable feeling" whenever he heard scriptural expressions in his later life and a repugnance for religious worship. Also, Spencer indicates that he later felt moral objections to the Christian doctrines of original sin and hell. Spencer felt no need for traditional religion and claimed that Christianity was "evidently alien to my nature, both emotional and intellectual." He did not experience a violent crisis of faith like so many of his contemporaries but, rather, he slowly and imperceptibly discarded Christianity during the late thirties and early forties. Spencer explained that "the current creed became more and more alien to the set of convictions formed in me, and slowly dropped away unawares. When the change took place it is impossible to say, for it was a change having no marked stages."[26]

In the middle fifties, before Mansel delivered his lectures, Spencer was sure "that the existence of a Deity can neither be proved nor disproved" (*LLHS*, 81). All of the evidence points to the conclusion that Spencer was well on his way toward becoming what we now call an agnostic before the late fifties brought Mansel or even Darwin to prominence. Spencer confirms this in a letter to F. Howard Collins of 1897: "My change from Theism to Agnosticism . . . took place long before the evolutionary philosophy was commenced, and long before I ever thought of writing it, and the change had nothing whatever to do with the doctrine of evolution. There has been no change whatever in that respect since 1860, when the writing of the philosophy was commenced" (*LLHS*, 398). Mansel's lectures were important for Spencer's intellectual development, but they did not cause him to plunge into a religious crisis.

Epistemology and Science before 1858

Prior to 1858, Spencer had already developed an elaborate epistemology. During the early and middle Victorian era, there were basically two prevalent philosophies of science founded on differing theories of knowledge. An empiricist epistemological standpoint was often found in conjunction with a philosophy of science which stressed the role of experience in the accumulation of scientific knowledge, while an intuitionist theory of knowledge was frequently linked with a philosophy of science which emphasized the importance of innate conceptions and German *Naturphilosophie* for knowledge of the physical world. Sir John Herschel summed up the situation well in 1841 in a review of Whewell's *Philosophy of the Inductive Sciences:*

> we are thus, at the very outset of the subject, presented with two Schools of such Philosophy—that which refers all our knowledge to *experience,* reserving to the mind only a high degree of activity and excursiveness in collecting, grouping, and systematizing its suggestions—and that which assumes the presence of *innate conceptions* and truths antecedent to experience, intertwined and ingrained in the very staple and essence of our intellectual being, and commanding, as with a divine voice, universal assent as soon as understood.[27]

Herschel went on to place Whewell in the latter group and himself in the empiricist tradition. Indeed, many scholars have considered Herschel and his *Preliminary Discourse on the Study of Natural Philosophy* (1830) as characteristic of the British empiricism espoused by Bacon, Newton, Locke, Hume, and J. S. Mill.[28] Whewell, however, is generally considered to be a "Kantian idealist," and his kinship with German thought caused his contemporaries to view his philosophy as essentially "un-British."[29]

Herbert Spencer seems to have affinities with both groups, but this turns out in the end to be deceptive. The other agnostics criticized him for being an a priori philosopher, and Huxley took great pleasure in teasing Spencer that his "idea of tragedy is a deduction killed by a fact."[30] In other words, Huxley believed that Spencer's love of theory often led him to ignore empirical fact. Spencer lent credence to this view when he attacked pure empiricism for failing to recognize that any primary assumptions required to build a philosophical system were in fact necessary, a priori truths.[31] However, although Spencer tended to treat empirical data as a means to illustrate, rather than test, his theory, he still belongs in the empiricist tradition. Spencer did borrow from the German thinkers, but he was not vitally interested in the Germano-Coleridgean mode of thought. He had a low opinion of Coleridge and referred to him as a mere plagiarist of Schelling's ideas (*PP,* 353).

Spencer took an equally dim view of Kant's so-called idealistic theory of knowledge. In his *Autobiography,* Spencer described how, in 1844, he first came into contact with a copy of a translation of Kant's *Critique of Pure Reason.* "This I commenced reading, but did not go far. The doctrine that Time and Space are 'nothing but' subjective forms,—pertain exclusively to consciousness and have nothing beyond consciousness answering to them,—I rejected at once and absolutely; and, having done so, went no further." Admitting that he was an impatient reader, Spencer made it a general rule never to continue reading a book if he dissented from its fundamental principles. Spencer then went on to criticize Kant for perversely contradicting "an immediate intuition of a simple and direct kind, which survives every effort to suppress it" (1:289).

This condemnation of Kant was repeated in Spencer's *Principles of Psychology* (1855). "That Space and Time are 'forms of sensibility' or 'subjective conditions of thought' that have no objective basis," Spencer remarked, "is as repugnant to common sense as any proposition that can be framed." Spencer held that our intuition of space as external is so clear and strong that the logical inference cannot be doubted. After his first encounter with Kant in 1844, Spencer asserted that "whenever, in later years, I have taken up Kant's *Critique of Pure Reason,* I have similarly stopped short after rejecting its primary proposition."[32]

Mansel fared no better than Kant, and was, in fact, dismissed by Spencer as a "Kantist" in *The Principles of Psychology* (56). It was through Mansel, as well as his friend G. H. Lewes, that Spencer acquired at least superficial knowledge of Kant.[33] Although Spencer came into contact with Kant indirectly, scholars have linked his agnosticism to the critical philosophy.[34]

Spencer's emphasis on humanity's direct perception of the external world seems to point to a similarity between his epistemological position and that of Hamilton's school of common sense. However,

Spencer insisted that Hamilton was wrong to assert that consciousness testifies to both the subject and object. In his eagerness to reject idealism, Spencer maintained that the consciousness of the object is much stronger than a consciousness of the subject. "Thus there is good ground for the belief that the cognition of the *non-ego* does not involve a simultaneous cognition of the ego-ground," Spencer declared, "which is strengthened by the remembrance that we can express cognition of objective being in words that involve no assertion of subjective being (the book exists), which we could not do did the one conception involve the other—and ground yet further strengthened by the consideration that we can perfectly well conceive an object to remain in existence after our own annihilation, *which it would be impossible to do if the cognition of the subject and object were simultaneous, and consequently inseparable.*" Spencer declined to accept Hamilton's natural realism and espoused a position he referred to as realism, or the belief "in objects as external independent entities" (*PP,* 47, 58).

It should be recalled that Kant referred to the transcendental realist as one who asserts that time and space are given in themselves independently of human sensibility, and that nature is a thing-in-itself or an entity existing independently of human beings. This would seem to be Spencer's viewpoint, for he asserts that objects "remain in existence after our own annihilation," which affirms that nature (appearances) is not related to us and that it would subsist even if we no longer existed. Spencer does not even indicate that he is aware of the crucial difference between maintaining that objects remain in existence after the *individual's* annihilation, and affirming that objects would remain in existence after the extinction of the *human race.* Kant would contend that one must hold to the former declaration while rejecting the latter statement in order to maintain the transcendental idealist position. Spencer blurs the distinction between these two statements because of his transcendental realism. Clearly, in Kantian terms, Spencer's "realism" is but another form of Kant's transcendental realism since Spencer sees time, space, and nature as independent, self-existing entities.

It is not surprising that Spencer the empiricist should turn out to be a transcendental realist. Most Victorian thinkers were implicit transcendental realists. Herschel conceived of space as "a substantive reality independent of our minds" and contrasted this to Whewell's Kantian notion of space as a condition or form. Yet it can be argued that even Whewell misunderstood Kant's notion of nature as relation, for, although the former held that experience is not produced by a passive reception of sensations but rather by an interpretative act of perception, he did not believe that nature is transformed by our knowing it. Dingle claims that Whewell, like any other Victorian thinker of the 1850s, assumed that the task of scientists was to study the world of material objects lying before them by "direct observation, by the use of instruments, and by experimental arrangement of conditions so as to facilitate observation and measurement" and that "none of these processes was conceived to change the world in any way." The prevailing philosophy conceived of correct Newtonian scientific method as a search for the universal causal laws that determine the course of events in a real external world. The situation was beautifully precise and clear-cut in its Cartesian duality. On the one side lay the world to be known and examined, the material or natural world, which stood off from the scientist and preserved an unalterable reality independent of the mind that observed it. On the other side were those minds that discovered what the material world contained and how it was ordered through the use of observation, experiment, and rational deduction. The Victorian scientist viewed nature as a self-existing, independent entity unrelated to human beings.[35]

Spencer did not believe that it was necessary to resort to Kant's transcendental idealism in order to understand why we are unable to banish the ideas of space and time from our minds. "Our powerlessness to conceive the non-existence of Space," Spencer claimed, "requires no such hypothesis as that of Kant for its explanation" because the "experience-hypothesis explains all that the Kantian hypothesis is intended to explain." To Spencer, space was not a form of thought but rather a form or quality of nature revealed to us through experience. "If space be an universal form of the *non-ego*," Spencer declared, "it must produce some corresponding universal form in the *ego*—a form which, as being the constant element of *all* impressions presented in experience, and therefore of *all* impressions represented in thought, is independent of every *particular* impression; and consequently remains when every particular impression is banished" (*PP*, 54, 230).

During the early fifties, when Spencer wrote *The Principles of Psychology*, he was working on a Lamarckian evolutionary theory that depended on adaptation to changing circumstances and the inheritance of acquired characteristics.[36] He explicitly admits his adherence to the "development hypothesis" in *The Principles of Psychology* and defines it as "the belief that Life under all its forms has arisen by a progressive, unbroken evolution; and through the immediate instrumentality of what we call natural causes." Spencer attempted to use the "development hypothesis" as further evidence that space and time were derived from human experience. The "development hypothesis," he claimed, "furnishes a solution of the controversy between the disciples of Locke and those of Kant" for "joined with this hypothesis, the simple universal law that the cohesion of psychical states is proportionate to the frequency with which they have followed one another in experience, requires but to be supplemented by the law that habitual psychical successions entail some hereditary tendency to such successions, which, under persistent conditions, will become cumulative in generation after generation, to supply an explanation of all psychological phenomena; and, among others, of the so-called 'forms of thought.'" In order to justify his rejection of Kant's transcendental idealism, Spencer conceived of time and space as real qualities inhering in "external" natural objects, and he combined this notion with his stress on evolutionary progress as people's ability to adjust continuously their internal relations (mind) to external relations (nature). "The manifestations of intelligence," Spencer submitted, "are universally found to consist in the establishment of correspondences between relations in the organism and relations in the environment; and the entire development of intelligence is seen to be nothing else than the progress of such correspondences in Space, in Time, in Speciality, in Generality, in Complexity" (*PP*, 482, 578).

Spencer differed from the other empiricists like Mill because he seriously attempted to understand the origin of elements of knowledge considered by intuitionists to be a priori. Through his use of the "development hypothesis," he was able to conceive of these a priori forms of thought as the product of race experiences that are a priori for the individual in the sense that they are given to him or her, but a posteriori for the entire series of individuals.[37] However, Spencer still considers these a priori elements of knowledge to be the product of experience and the law of association (*PP*, 526). Spencer's conception of a priori forms of thought is actually more physiological than epistemological, and this once again underlies the vast difference between Spencer and Kant. "In the sense, then, that there exist in the nervous system certain pre-established relations answering to relations in the environment," Spencer asserted, "there is truth in the doctrine of 'forms of thought'—not the truth for which its advocates contend, but a parallel truth. Corresponding to absolute external relations, there are developed in the nervous system absolute internal relations—relations that are developed before birth; that are antecedent to, and independent of, individual experiences; and that are automatically established along with the very first cognitions" (*PP*, 583).

Spencer often made attacks on the pure empiricist standpoint, as he denied that knowledge is derived wholly from the experiences of the individual. He saw himself as reconciling the theories of the pure empiricists and the transcendentalists, "neither of which is tenable by itself" (*PP*, 580). It is not difficult to misconstrue Spencer's position and see him as an a priori philosopher. Mill actually made the mistake of grouping Spencer with other a priori philosophers such as Hamilton, Whewell, and Kant in his *Examination of Sir William Hamilton's Philosophy.* "I am taken aback at finding myself classed as in the above paragraph," Spencer protested, "considering that I have endeavored to show how all our conceptions, even down to those of Space and Time, are 'acquired'—considering that I have sought to interpret forms of thought (and by implication all intuitions) as products of organized and inherited experiences."[38]

Launching the Synthetic Philosophy

By 1858, when Mansel delivered his Bampton Lectures, Spencer had already lost his faith and developed an empiricist philosophy of science. However, he had not yet elaborated the connection between his epistemology and his philosophy of religion. His first statement on this issue was unveiled in Part One of *First Principles* (1862), "The Unknowable." *First Principles of a New System of Philosophy* was intended as the initial volume of Spencer's whole system of philosophy, and he encountered a great deal of difficulty getting his project off the ground. A multi-volume scientific synthesis was not considered commercially promising, and Spencer searched in vain for a job that would support him as he wrote. Then Spencer thought of a scheme whereby he would publish the work in installments appended to each number of the *Westminster Review* for which he would be paid regularly, but the plan fell through.[39] In the autumn of 1859 Spencer hit upon the idea of issuing the system in a serial form to subscribers who would each pay 10s. yearly.[40] From the income thereby received Spencer could establish an independent financial base from which to begin his ambitious project.

On 29 March 1860 Spencer issued a comprehensive program for his proposed system of philosophy. "The prospectus contained a general sketch of the scheme," Spencer reports, "the successive volumes being described; their divisions into parts; and the natures of the contents of these parts. The conception had at that time been so far developed in its general outlines that no deviation from the prospectus has been found needful in the course of execution—the divisions and subdivisions have been successively published in the order and form originally specified."[41] Spencer's pride in the fact that the outline and conceptions of his entire program were never altered from their original statement has been viewed by others as a defect in his philosophy. It is a striking irony that a system whose content is evolution should be presented in a static form. But Spencer did base his whole system on the idea of evolution while simultaneously preventing his system from ever evolving.[42]

First Principles was begun in 1860, the first installment being issued in October. Thereafter, successive numbers appeared until June 1862, when the book was finished and published as a volume, despite Spencer's bouts of nervous exhaustion and insomnia.[43] Spencer's *First Principles* was both the first major statement of agnosticism and the most comprehensive account of its basic tenets. Although Huxley had not yet coined the term *agnosticism* when *First Principles* was published, Spencer later accepted the term as an appropriate designation for his convictions, and Victorian thinkers regarded *First Principles* as the agnostic Bible. Despite Spencer's earlier antipathy toward Kant and Mansel, he found *The Limits of Religious Thought* quite useful for his work in *First Principles.* If it can be demonstrated that Spencer's work in this volume was significantly influenced by Mansel, then we are at least halfway to validating the thesis that the Kantian tradition is important for our understanding of agnosticism.

Spencer's avowed purpose in "The Unknowable" was to reconcile science and religion. He begins by pointing out that there is usually something held in common even between opposite beliefs. Spencer proposed to use this observation as a method for determining truth. "This method is to compare all opinions of the same genus," he stated, "to set aside as more or less discrediting one another those various special and concrete elements in which such opinions disagree; to observe what remains after the discordant constituents have been eliminated; and to find for this remaining constituent that abstract expression which holds true throughout its divergent modifications" (11). Spencer concluded that we must use this principle in searching for a common element that can reconcile science and religion: "Since these two great realities are constituents of the same mind, and respond to different aspects of the same Universe, there must be a fundamental harmony between them" (24). He warned, however, that the common element of unity would be an abstract principle, for science could not be expected to recognize special religious doctrines such as the Trinity, while religion could take no cognizance of special scientific doctrines.

The Contradiction of Ultimate Ideas

The second chapter of "The Unknowable" was taken up with an analysis of "Ultimate Religious Ideas." Spencer began by discussing "the formation of symbolic conceptions, which inevitably arises as we pass from small and concrete objects to large and to discrete ones," and is "mostly a very useful, and indeed necessary, process" (27). This process of symbolization is the only way we can deal with heterogeneous objects or things that possess a vast number of attributes. We flatten them out, select a symbol that omits some attributes and, therefore, are left with an inadequate representation of such objects. Spencer points out that the process of symbolization is dangerous, for we can "habitually mistake our symbolic conceptions for real ones" (27).

Spencer then moved on to a discussion of the three different suppositions respecting the origin of the universe—atheism, pantheism, and theism. Mansel had examined each of these positions and found that reason was unable to justify any of them; Spencer's argument is almost identical. Atheism or the self-existence of the universe, implies to Spencer a notion of that which has no beginning. Spencer dismisses this as inconceivable, unintelligible, and irrational. Pantheism, or self-creation, is as inconceivable as atheism for "to conceive self-creation, is to conceive potential existence passing into actual existence by some inherent necessity" (32). Finally, theism, or creation by an external agency, cannot be justified by reason, because to account for it "only the same three hypotheses are possible—self-existence, self-creation, and creation by external agency" (35). Spencer asserts that the three hypotheses do not stand for real thought but merely suggest vague symbols, and he thus links his comments on the formation of symbols to atheism, pantheism, and theism. All three "involve symbolic conceptions of the illegitimate and illusive kind" (36).

The contradictions in each hypothesis illustrate, for both Mansel and Spencer, reason's inability to deal with the problem. "Thus these three different suppositions respecting the origin of things," Spencer wrote, "verbally intelligible though they are, and severally seeming to their respective adherents quite rational, turn out, when critically examined, to be literally unthinkable" (35). Spencer was able to utilize Mansel's argument in *The Limits of Religious Thought* as support for his chapter on "Ultimate Religious Ideas." He not only modeled his strategy on Mansel's position, but also lifted pages of quotes from the Bampton Lectures dealing with the contradictions inherent in viewing the absolute, the infinite, and the first cause in conjunction as attributes of the same being to prove his contention that the concept of an absolute and infinite first cause was an illegitimate symbol. "Here I cannot do better than avail myself of the demonstration which Mr. Mansel," Spencer announced, "carrying out in detail the doctrine of Sir William Hamilton, has given in his 'Limits of Religious Thought.' And I gladly do this, not only because his mode of presentation cannot be improved, but also because, writing as he does in defence of the current Theology, his reasoning will be the more acceptable to the majority of readers" (39).

The Bampton Lectures created an intellectual atmosphere quite advantageous to the aims of budding agnostics. Any fears that they would shock the Victorian public were dispelled by the fact that they espoused an epistemological position already discussed by a member of the establishment. The chapter on "Ultimate Religious Ideas" is concluded by the assertion that all religions share the ultimate religious truth that there is a mystery to be solved. Spencer rejected the different solutions to this mystery (such as atheism, pantheism, and theism) and went on to find the most basic level of agreement between them.

The next chapter of "The Unknowable" contains a discussion of "Ultimate Scientific Ideas." Since Spencer's aim is to reconcile science and religion, he finds that ultimate scientific ideas, like ultimate religious ideas, are all "representative of realities that cannot be comprehended" (66). In rapid succession, Spencer demonstrates the contradictions inherent in ultimate scientific ideas of space, time, matter, motion, and force. This was an extension of Mansel's attack on rational theology into the realm of scientific thought. In the case of time and space, Spencer asserted that they cannot be conceived as either entities, the attributes of entities, or nonentities. He continued on this theme by repeating his rejection of Kant as expressed in the earlier *Principles of Psychology* (49). Spencer reiterated his transcendental realism but admitted, for the sake of his argument, that though we have an "insurmountable" belief in the objective reality of time and space due to our immediate knowledge of them, "we are unable to give any rational account of it" (50). Equally incomprehensible is the ultimate nature of matter. Spencer reached this conclusion after discussing our inability to conceive of matter as either infinitely divisible or finitely divisible, absolutely solid, or composed of atoms.

The fourth chapter of "The Unknowable" is entitled "The Relativity of All Knowledge." Whereas Spencer's earlier chapters resemble Mansel's rejection of the ontological approach to a philosophy of religion, this section parallels Mansel's view of the psychological method. "The demonstration of the necessarily relative character of our knowledge," Spencer claimed, "as deduced from the nature of intelligence, has been brought to its most definite shape by Sir William Hamilton" (74). Spencer therefore supported his contention that absolute knowledge is beyond us by quoting from Hamilton's "Philosophy of the Unconditioned." However, Spencer's argument developed into a rehash and quotation of Mansel's theory in *The Limits of Religious Thought* that every complete act of consciousness implies distinction and relation (82). Just as Mansel had pointed to the psychological approach as an explanation for the failure of the ontological approach, Spencer stated that "we not only learn by the frustration of all our efforts, that the reality underlying appearances is totally for ever inconceivable by us; but we also learn why, from the very nature of our intelligence, it must be so" (98).

The final chapter of "The Unknowable" discussed the reconciliation between science and religion. Although the burden of the previous chapter concerned the unknowable nature of the thing-in-itself, Spencer insisted that we retain a consciousness of the actuality lying behind appearances, which in turn explains our indestructible belief in that actuality. "At the same time that by the laws of thought we are rigorously prevented from forming a conception of absolute existence," Spencer asserted, "we are by the laws of thought equally prevented from ridding ourselves of the consciousness of absolute existence" (96). This consciousness of the absolute was described by Spencer as indefinite, yet positive. He maintained that, though it was impossible "to give this consciousness any qualitative or quantitative expression whatever, it is not the less certain that it remains with us as a positive and indestructible element of thought" (91).

Spencer's agnosticism may strike us as being peculiar because he agreed with Mansel that an intellectual limit implied the existence of something beyond. Mansel asserted that "the existence of a limit to our powers of thought is manifested by the consciousness of *contradiction*, which implies at the same time an attempt to think and an inability to accomplish that attempt." If we stop here, the position articulated is pure neutral agnosticism. But Mansel continued by adding that "a limit is necessarily conceived as a relation between something within and something without itself; and thus the consciousness of a limit of thought implies, though it does not directly present to us, the existence of something of which we do not and cannot think" (*LRT*, 62). This positive view of the significance of intellectual limitation is also part of Spencer's standpoint.

In fact, Spencer took Mansel to task for not stressing strongly enough that human beings possess a positive consciousness of the absolute. During a later controversy with the Positivist Frederic Harrison, Spencer reiterated the point. "For whereas," Spencer declared, "in common with his teacher Sir William Hamilton, Dean Mansel alleged that our consciousness of the Absolute is merely 'a negation of conceivability;' I have, over a space of ten pages, contended that our consciousness of the Absolute is not negative but positive, and is the one indestructible element of consciousness 'which persists at all times, under all circumstances, and cannot cease until consciousness ceases.'"

Spencer went on to distinguish between "Comtean Agnosticism which says that 'Theology and ontology alike end in the Everlasting No with which science confronts all their assertions,'" and his brand of agnosticism "set forth in *First Principles*, which, along with its denials, emphatically utters an Everlasting Yes."[44]

It is this positive quality of Spencer's agnosticism which, to him, is the basis of the reconciliation between science and religion, for both pointed to a mysterious power underlying phenomena. "Common sense asserts the existence of a reality," Spencer affirmed, "Objective Science proves that this reality cannot be what we think it; Subjective Science shows why we cannot think of it as it is, and yet are compelled to think of it as existing; and in this assertion of a Reality utterly inscrutable in nature, Religion finds an assertion essentially coinciding with her own" (*FPNSP*, 99). This, then, is the abstract element that both science and religion hold in common and for which Spencer has been searching throughout "The Unknowable."

Religion and the Unknowable

If we compare *First Principles* with *Principles of Psychology*, both of them epistemological works, we can perceive the debt Spencer owed to Mansel. There is no notion in *The Principles of Psychology* that ultimate scientific ideas are plagued by contradictions. Space and time seem to be completely intelligible on the experience hypothesis. There is nothing in the earlier book vaguely resembling the stress on Hamilton's "philosophy of the conditioned" in *First Principles*. All of this points to the conclusion that Spencer saw in Mansel's *Limits of Religious Thought* a number of arguments that he could easily adapt for his own use. Spencer undoubtedly borrowed from Mansel the strategy of building antinomies in order to demonstrate the impotence of reason in religious affairs and the emphasis on relation and distinction as conditions of thought. Although it would be an exaggeration to claim that Spencer's personal values were profoundly transformed by the Bampton Lectures, Mansel did supply him with the tools for building an elaborate agnostic epistemology that was meant to be the foundation stone for the whole Synthetic Philosophy.

Spencer's use of Manselian epistemology to ground his reconciliation of science and religion has been criticized for a number of reasons. The major complaint has been that, in placing science in the realm of the knowable and religion in the world of the unknowable, Spencer's reconciliation was false, unsuccessful, and inconsistent. Science suffers in that ultimate scientific ideas cannot be rationally established or even established as rational. Indeed, Spencer's position appears to be unusually self-destructive in that *First Principles* proposes to base a whole system of knowledge on inconceivable ideas.[45]

More common is the charge that Spencer's reconciliation is brought about at the cost of the power of religion. Spencer believed that his reconciliation protected both science and religion by restricting them to their proper spheres of influence:

> Gradually as the limits of possible cognition are established, the causes of conflict will diminish. And a permanent peace will be reached when Science becomes fully convinced that its explanations are proximate and relative; while Religion becomes fully convinced that the mystery it contemplates is ultimate and absolute. Religion and Science are therefore necessary correlatives. As already hinted, they stand respectively for those two antithetical modes of consciousness which cannot exist asunder. A known cannot be thought of apart from an unknown. (*FPNSP*, 107)

However, though safe from attack by science, religion had lost much in Spencer's scheme of things. Revelation, which Mansel used to build a new philosophy of religion when he reached the limits of thought and knowledge, had no place in Spencer's thought. Neither did Mansel's stress on personal communion with God. In fact, Spencer would not allow for any type of theology, be it derived from a revised psychological approach or not. Furthermore, Spencer protested Mansel's jump from the unknowableness of God to God as person (*FPNSP*, 108). Sheldon charged that Spencer's plan achieved a reconciliation only because "if the program should be strictly carried out, there would not be enough of religion left to seriously antagonize science or anything else."[46]

While religion lost the personality of God and revelation in Spencer's reconciliation, science had everything to gain. Although science had to admit to the existence of the Unknowable, it was now free to explain the world purely in terms of matter and motion. Religion was relegated to the sphere of the Unknowable, Cockshut maintains, so that the study of the knowable might proceed unhampered.[47]

Spencer's resolution of the conflict between science and religion seems so one-sided in setting such favorable terms for science that a number of scholars have seriously doubted the sincerity of Spencer's avowed religious sentiments. Copleston denies that Spencer's Unknowable represents "a genuinely religious element" and declares that it is a mistake to compare it "with the Christian doctrine of God's incomprehensibility." Metz echoes Copleston's point in his declaration that Spencer's Unknowable is in reality "only a decoration of the façade, intended to give to the structure an appearance less repellent to religious minds." Sheldon views Spencer's system as materialist and "antitheistic in tenor."[48] Furthermore, many have seen Spencer's affirmation of the existence of the Unknowable to be a blatant contradiction of the very premises of his thought, and they therefore feel justified in questioning the validity of Spencer's theism. Spencer's inconsistency stems from a scepticism that is so destructive that he cannot affirm anything positive about God, including his existence.[49] Even fellow unbelievers were somewhat leery of Spencer's theism. J. A. Froude, upon receiving the prospectus for the whole project, which included a description of *First Principles,* was puzzled by Spencer's aim. "Mansel says the absolute is the unknowable," Froude wrote to Spencer in 1860. "How by following all *his* reasonings you are to establish a belief in it, I am curious to see" (*LLHS*, 97).

But despite Spencer's attempt to remove from religion a great deal of Christian doctrine, and despite his apparent inconsistency in affirming the existence of a supreme being, both Spencer's genuine religiousness and his reverence for the Unknowable must be recognized. It is close to impossible to prove the religious sincerity of any thinker, particularly in the case of a cerebral and unemotional figure like Spencer, who had a reputation for heterodoxy. However, there seems to be no good reason for doubting his public statements on his religious sentiment and belief in the existence of the Unknowable, since they are confirmed by private letters to friends and are important components of his whole system.

In his *Autobiography* Spencer recalled his determination to begin his grand project with a preface that would set forth his views on ultimate metaphysical and theological questions and would thereby remove any suspicions that he was presenting a materialist philosophy. "My expectation," Spencer wrote, "was that having duly recognized this repudiation of materialism, joined with the assertion that any explanation which may be reached of the order of phenomena as manifested to us throughout the Universe, must leave the Ultimate Mystery unsolved, readers, and by implication critics, would go on to consider the explanation proposed."[50] Spencer hoped that critics would accept his sincere claim that conceiving of God as the Unknowable was not a disguise for his materialism or atheism but in fact represented a sophisticated version of theism. To Spencer, the choice was not between a notion of personality (whether regulative or not) and something lower, but rather "between personality and something higher," for he believed that it was possible for there to be a mode of existence that entirely transcended what he considered the anthropomorphic conception of being. Assigning attributes to God that are derived from human nature, such as personality, in reality degrades God in Spencer's eyes. Religion is barred from possessing any knowledge of God, and Spencer calls any attempt to know God impious (*FPNSP,* 109–10). Spencer's God is thus a completely abstract, impersonal, unknowable entity that gives life and reality to appearances. He was usually very careful about how he talked about God, because to make any assertions regarding God's nature conflicts with Spencer's anti-anthropomorphic agnosticism. However, he sometimes used terms drawn from nature, such as "Infinite and Eternal Energy" to describe his deity.[51]

On 9 November 1882 John Fiske spoke on "Evolution and Religion" at the farewell dinner held in New York in honor of Herbert Spencer's trip to the United States. Fiske became one of Spencer's chief spokesmen on the theological left wing of American Protestantism in the sixties when he turned against the Calvinist theology of his youth and embraced the new Spencerian philosophy. In his speech Fiske declared that "Mr. Spencer's work on the side of religion will be seen to be no less important than his work on the side of science, when once its religious implications shall have been fully and consistently unfolded." For Fiske, Spencer's evolutionary system asserted, "as the widest and deepest truth which the study of nature can disclose to us, that there exists a Power to which no limit in time or space is conceivable, whether they be what we call material or what we call spiritual phenomena, are manifestations of this infinite and eternal Power. Now, this assertion, which Mr. Spencer has so elaborately set forth as a scientific truth,—nay, as the ultimate truth of science, as the truth upon which the whole structure of human knowledge philosophically rests,—this assertion is identical with the assertion of an eternal Power, not ourselves, that forms the speculative basis of all religions."[52]

Spencer was delighted with Fiske's speech, and when he returned to England he wrote a letter, dated 24 November 1882, which emphasized the significance of Fiske's theme. "I wanted to say how successful and how important I thought was your presentation of the dual aspect, theological and ethical, of the Evolution doctrine," Spencer declared. "It is above all things needful that the people should be impressed with the truth that the philosophy offered to them does not necessitate a divorce from their inherited conceptions concerning religion and morality, but merely a purification and exaltation of them."[53]

The religious and theological dimension was absolutely essential to Spencer's whole evolutionary scheme. It was the existence of the Unknowable which guaranteed that beneath the seeming waste of the evolutionary process lay an economy, order, purpose, and harmony. Spencer's life was devoted to the attempt to prove scientifically that his faith in the benevolence of natural processes was not misplaced.

Beatrice Webb reported that by the time she came to know Spencer his "first principles" had become "a highly developed dogmatic creed with regard to the evolution of life. What remained to be done was to prove by innumerable illustrations how these principles or 'laws' explained the whole of the processes of nature, from the formation of a crystal to the working of the party system within a democratic state."[54] Spencer's enterprise was as much religious as scientific, and in his system Mansel's epistemology became the basis of a brand of agnosticism raised to the position of a theistic concept.

Spencer's *First Principles* appeared at a critical point in the intellectual development of Huxley, Tyndall, Stephen, and Clifford. By effectively using Mansel for his own ends, Spencer showed the others how the notion of the limits of knowledge could be turned against Christian orthodoxy. Mansel's approach to epistemology became enshrined into the very heart of agnostic thought.

Chapter Four

DISILLUSIONMENT WITH AND ATTACK ON ORTHODOXY

> *May you not say, in language strong enough to satisfy a Positivist, that the human mind can form no conception of Divinity; that good and merciful, applied to the Almighty, mean no more than wrathful and jealous, or even than epithets implying corporeal attributes, and say it all amidst general applause so long as your assault is ostensibly directed against the presumptuous Deist, and not against Moses or St. Paul? A grateful clergy will applaud you for wielding weapons so unfamiliar to them, and so steadily associated with the adversary, and will take your word for it that you mean well.*
>
> LESLIE STEPHEN

The agnostics delighted in exploiting the utterances of eminent Christian divines for their own use. The defense of scientific naturalism led Huxley, Stephen, Tyndall, and Clifford to seize upon Mansel's notion of the limits of knowledge as well as themes dealing with the mysteriousness of God drawn from the writings of fideists. Their sensitivity to the sceptical quality of Christian theology was the outcome of many years of reflection, religious questioning, and ultimately, a disillusionment with Christian orthodoxy, during a time when the old Anglican-aristocratic order was being challenged by those who envisioned a new social order.

The Social Significance of Scientific Naturalism

Frank Turner has argued that the conflict between science and religion in Victorian England was more than a dispute over ideas, for it reflected the collision between established and emerging intellectual and social elites vying for popular cultural preeminence in a modern industrial society.[1] Similarly, the agnostics' interest in using Mansel's theory of knowledge in their war with the Church can be seen in connection to a crucial social dimension of their activity. The precarious place of science in mid-Victorian society made some budding young scientific naturalists prepared to listen closely to Mansel's sceptical arguments and adapt them for their own cause. When Huxley and Tyndall were searching for posts from which to build their scientific careers in the early fifties they had to endure years of frustration before they found suitable positions. Like other members of the developing new elite the major obstacles to their success were their humble middle-class origin, their lack of contact with the major established English intellectual institutions, and the institutional structure of English society, which did not allow for the existence of many careers for professional scientists. The resources of the academic profession were organized to suit the interests of the Anglican Church and the Tory party. Through its control of the universities, the Church had a virtual monopoly over potential funds for the development of a scientific professoriate. Simultaneously, the Church destroyed any justification for the existence of such a professoriate by organizing school curricula in a fashion bound to lower the number of scientifically trained students moving into higher education.

In order to carve out a place for themselves in Victorian social and intellectual life, and to secure their future, Huxley, Tyndall, and the other scientific naturalists embarked on a program to professionalize science by obtaining from industry, government, and education the resources needed for salaries and research facilities. But to achieve this goal they were brought into conflict both with those amateurs and parson naturalists inside science who did not share their aims and with those outside the profession who resisted science's claims of self-definition. The attempt to professionalize science involved a desire for social and occupational, as well as intellectual, independence, because scientific naturalists argued for an empirically based discipline that need not take into account the Bible, the doctrines of the Anglican Church or the opinions of the clergy. As an ideology scientific naturalism served the interests of sections of the new professional middle class and provided a rationale for their leaders to wrest cultural and social prestige from the clergy. Mansel's epistemology supplied scientific naturalists with powerful fuel for shaking the authority of the Church. The notion of the limits of knowledge was a useful tool in defining the boundaries of proper science and attacking both those within and those outside the profession who were determined to go beyond these boundaries by bringing improper theological concepts into science.

The Chronology of Agnosticism

Up to this point we have examined Mansel's *Limits of Religious Thought*, published in 1858, and the appearance of Spencer's *First Principles* in 1862 amid a larger controversy surrounding the issue of human knowledge of God. During the sixties, when Huxley, Stephen, Tyndall, and Clifford were trying to develop a new creed as they worked out the larger implications of evolutionary theory and biblical criticism, the Mansel controversy molded their thought patterns. They began to reach some tentative conclusions by the end of the decade, and they spoke out publicly on their agnostic beliefs during the seventies.

The coining of the term *agnosticism* in 1869 by Huxley was the prelude to the decade during which agnosticism, as a body of theory, reached its fullest development. Huxley's *Lay Sermons, Addresses, and Reviews* was published in 1870, followed by Stephen's *Essays on Freethinking and Plainspeaking* (1873). Tyndall's "Belfast Address," which provoked as much controversy as any of Huxley's essays or the Wilberforce-Huxley showdown at Oxford, was delivered in 1874.[2] The publication of Stephen's "An Agnostic's Apology" (1876) marked the first time an important agnostic actually used the term *agnostic* in print and defended its validity as an appropriate response to the bankruptcy of Christian orthodoxy.[3] It was not until 1878 and the publication of *Hume* that Huxley initially mentioned his coinage *agnosticism* and briefly discussed the meaning of the word in relation to the philosophy of Hume and Kant (58). Throughout the seventies Clifford had been publishing a number of controversial essays that were gathered together in *Lectures and Essays* (1879) by Leslie Stephen and Frederick Pollock after Clifford's death. The outspoken nature of such essays as "The Ethics of Belief" (1877) and "The Ethics of Religion" (1877) earned Clifford the nickname "that delicious *enfant terrible*" from William James.[4]

The codification of agnostic theory in the seventies was partly the result of the growth of personal friendships between agnostics as well as the organization of societies wherein they could discuss issues of mutual concern and further refine their views. The major agnostics never formed an organized school or sect, but they regarded each other as friends and shared a common circle of acquaintances, quoted one another with approval in their writings, and lent support, both moral and financial, in times of need.

Huxley and Tyndall were the first to meet, in 1851, at a British Association meeting at Ipswich. At the time both were attempting to secure a permanent position in the scientific community. Their common struggles drew them closer together, and they affectionately referred to each other as brother. Due to their friendship they were closely identified in the public mind, to the point where Tyndall's marriage to a titled lady led people to treat Mrs. Huxley more deferentially, and Tyndall's death resulted in reports on services held in Huxley's memory.[5]

Huxley, who had been introduced to Spencer in 1852, brought together his two friends, Tyndall and Spencer, a year later in the rooms of the Royal Society. "There commenced," Spencer recalled in 1894, "one of those friendships which enter into the fabric of life and leave their marks."[6]

In 1864, Spencer, Huxley, and Tyndall helped found the X-Club, which consisted of a set of mutual friends who were to become prominent in their respective fields of research, including William Spottiswoode (mathematics), Joseph Dalton Hooker (botany), Edward Frankland (chemistry), John Lubbock (archaeology), Thomas Archer Hirst (mathematics), and George Busk (medicine, zoology, paleontology).[7] The X-Club was a private, informal society where the members could exchange ideas on literature, politics, and science over dinner. For twenty years the members of the club met once a month from October to June. In his journal Hirst reported in 1864 that besides personal friendships "the bond that united us was devotion to science, pure and free, untrammelled by religious dogmas. Amongst ourselves there is perfect outspokenness, and no doubt opportunities will arise when concerted action on our part may be of service."[8] Hirst's remarks were prophetic, as the X-Club wielded tremendous power in the scientific world. Barton has likened it to the cabinet of a liberal party in science.[9] The formation of the club allowed the members to pursue a number of common objectives, including the advancement of research, the diffusion of science, and the reform of the public image of science. The X-Club provided a meeting place for like-minded men to discuss ideas and projects relevant to the development of agnosticism.

The Metaphysical Society, founded in 1869, also brought the agnostics closer together and impelled them toward formulating their creed. Spencer declined to participate, but both Tyndall and Huxley were members from the beginning, while Clifford joined in 1874 and Stephen in 1877. It was during the seventies that the four of them really came to know each other. Stephen, an admirer of "plainspeaking," wrote to Clifford in 1879 that "there is always something refreshing to one's soul in Huxley's writing—none of your shuffling and equivocating and application of top-colour." Stephen also gleefully related to Clifford an incident at a meeting of the Metaphysical Society where Huxley "trod rather heavily upon Sidgwick's toes" (*LLLS*, 333). In 1894, Stephen wrote to Huxley "I thoroughly enjoy reading whatever you write," complimented Huxley on his literary abilities, and declared "I agree with the substance of nearly all that you say" (ICST-HP 27:66). Stephen valued Huxley's friendship greatly and drew closer to him later in life.[10] Tyndall was also regarded by Stephen as "a very friendly 'acquaintance' too, enough to be called a friend."[11] Of Stephen, Tyndall wrote in 1893, "I consider his to be one of the firmest and most penetrating intellects of our day."[12] But although Stephen liked Huxley and Tyndall and referred to them as "some of the best as well as ablest men of the time," he was never part of their intimate circle of friends.[13]

Clifford, the youngest of all the agnostics, was just beginning to make a name for himself in the late sixties. Stephen, Huxley, and Tyndall all enjoyed his friendship and held a very high opinion of his scientific capabilities. Stephen described Clifford as "a real man of genius" (*LLLS*, 336). In 1890, shortly before his death, Tyndall recalled visiting the "gifted Clifford" eleven years earlier (*NF*, 377). Huxley, Tyndall, and Stephen had high hopes for an organization that Clifford had helped to found in 1878, the short-lived Congress of Liberal Thinkers. With Huxley as president, the congress was meant to bring together leading men from all parts of the United Kingdom, Europe, and America who were interested in liberating humanity from degrading dogma.[14] For a brief period of time this organization seemed to put Stephen, Clifford, Huxley, and Tyndall into closer contact.[15] But since Clifford was the driving force behind the congress, his death in 1879 spelled the end of the organization.

During the late seventies Clifford was slowly wasting away of consumption. Stephen, Huxley, and Tyndall did all they could to rally their friend. Tyndall and Huxley were involved in Stephen's scheme to raise a subscription to send Clifford to Madeira in the hope that the climate would help him recover. All three visited Clifford to keep up his spirits. After one visit Huxley left Clifford's house and exclaimed, his face clouded with despair, that "the finest scientific mind born in England for fifty years is dying in that house."[16] Stephen was particularly attentive. Clifford's widow appreciatively recalled Stephen's frequent visits, usually twice a week, in the last years of Clifford's life.[17] After Clifford's death Stephen continued to care for his widow and family, edited Clifford's philosophical papers, and later wrote the *Dictionary of National Biography* article on Clifford's life.

The Loss of Faith

Before the sixties and seventies the original agnostics were groping their way toward the articulation of a new religion without the benefit of any formal organization or intellectual signposts. Sometime during their lives Huxley, Stephen, Tyndall, and Clifford lost their faith in the Christianity of their time. It will help us to understand the role of Spencer, Mansel, and the Kantian tradition, in this process of alienation from Victorian Christianity, if we examine the crisis of faith experienced by each agnostic.

Huxley seems to have drifted slowly into unbelief. In 1840, at the age of fifteen, he supported the disestablishment of the Church of England and no longer accepted evangelical theology.[18] Working in the early forties as a medical assistant in the east end of London and seeing for himself the miserable plight of the working class, Huxley became more receptive to Carlyle's indignant attacks on complacent Victorian society. Carlyle's message inspired Huxley with a passion for social reform and a disdain for conventional Christianity. He also had been reading Hamilton, the fine points of which no doubt were beyond a teenager, but Huxley managed to grasp the essence of the Scottish philosopher's message. In 1860, he wrote to Kingsley that "it must be twenty years since, a boy, I read Hamilton's essay on the unconditioned, and from that time to this, ontological speculation has been a folly to me" (*LLTHH* 1:234). In his review of Arthur Balfour's *Foundations of Belief* in 1895, Huxley quoted the section in Hamilton's "Philosophy of the Unconditioned" concerning a "learned ignorance" as the consummation of knowledge, and asserted that the passage represented "the original spring of Agnosticism." "Here is the cardinal proposition of Agnosticism," Huxley affirmed, "as I understand it, set forth, with a force and clearness that have never been surpassed, sixty-six years ago."[19] However, a reading of Hamilton did not fill the vacuum left by his rejection of a Christian worldview.

Huxley was still seeking an alternative to Christianity when he embarked on his voyage on the *Rattlesnake* in 1846. His *Thätige Skepsis*, or active doubt, gave him no solid ground to stand upon.[20] In his journal he pointed to the advantages afforded by the distractions of a sea voyage. It allowed him to "get rid of the 'malady of thought'" that plagued him, the pain that he experienced in grappling with the chaotic state of his religious emotions.[21] However, by 1850 Huxley had worked out the basis of his agnostic creed. "Thanks to Hamilton and Mill," Huxley claimed, "the fundamental principles of what is now understood as Agnosticism were clearly fixed in my mind when, in 1850, I returned to England with a well-studied copy of Mill's *Logic*."[22] When appointed to the Royal School of Mines in 1854, and throughout the rest of the fifties, Huxley still believed in the value of religion, the eventual emergence of a new religion, and the existence of a God.[23]

Largely owing to the beneficial influence of Carlyle, his crisis of faith was limited to a questioning of the validity of Christianity.

In 1855 Tyndall wrote a letter to Huxley remarking upon "the great similarity of thought between us." "When I stand alone in the woods and hear the birds chirping, and see the trees sprouting I feel like a puzzled infant amid things which baffle my comprehension," Tyndall told Huxley, "I like to hear a man who instead of turning my stomach with dry theories of this universe is able to appreciate the difficulty of the problem and to recognize the fluxional character of our knowledge regarding it. The dogmatist however gets astride his little arc and swears it is the whole 360 degrees—We know better" (ICST-HP 8:23). Tyndall's spiritual development does bear resemblances to Huxley's religious evolution. Both were looking for a replacement for Christianity during the forties and fifties, both were attracted to Carlyle, and both retained a sense of the importance of religion.

Tyndall was born on 2 August 1820 at Leighlinbridge, County Carlow, in southern Ireland. His father, unsuccessful as a shoemaker and leather dealer, joined the Irish Constabulary. He was an ardent Orangeman who insisted on a strong Protestant household and educated his son in the art of theological debate.[24] At the age of eighteen Tyndall joined the Irish Ordinance Survey as a civil assistant. His work as surveyor and draftsman for the Irish Ordinance Survey took him to Youghal, Kinsale, and Cork, and in 1842 he was transferred to the English Survey in Preston. There he became a member of the Preston Mechanics' Institute in order to attend its lectures, use the library, and continue his program of self-improvement.

In 1843 Tyndall saw for the first time some extracts from Carlyle's *Past and Present* in the Preston newspapers. "I chanced," Tyndall remembered, "indeed, to be an eye-witness of the misery which at that time so profoundly moved Carlyle" (*NF*, 348). In his position as surveyor, Tyndall experienced firsthand the hard times of the forties, when riots broke out among the starving weavers in Preston. Social problems, Chartist unrest, and Carlyle's writings moved him toward a more radical political position. Unhappy with the inefficiency of the survey's administration and their unfairness to the Irish assistants, Tyndall formally protested and was subsequently dismissed in November 1843. The refusal of the Master General of the Ordinance (then Robert Peel) to see him about the matter, and poor job prospects, led Tyndall to seriously consider emigrating to America. However, in July 1844 a position was found in a private surveyor's office in Preston, and for the next three years Tyndall found himself in the thick of the railway mania of the late forties.

John Tyndall
A Drawing of 1860

By 1847 the railway boom was coming to an end, and Tyndall decided to move to another profession by accepting an appointment as teacher of mathematics at Queenwood College in Hampshire.[25] Fascinated by natural science, Tyndall went to Germany in 1848 to attend Marburg University, where he earned his doctoral degree from the Philosophical Faculty. Here he worked with the chemist Robert Bunsen and the physicist Hermann Knoblauch while establishing lifelong ties to the German scientific community. Returning to England in 1851, Tyndall was forced to take up his old position at Queenwood because he was unable to secure a professional post in the field of natural science. A number of years of frustration ended in 1853 when he was elected Professor of Natural Philosophy of the Royal Institution.

Tyndall's struggle to overcome the obstacles presented by his humble origins in order to force his way into the ranks of respectable society, and then the scientific world, left indelible marks on his spirit. He slowly discarded traditional Christianity and followed Carlyle in a religious radicalism and a political radicalism which were closely interconnected.

An entry in Tyndall's journal for 26 June 1847 reveals that by then he had given up many of the main doctrines of Victorian Christianity. "I cannot for an instant imagine," Tyndall wrote, "that a good and merciful God would ever make our eternal salvation depend upon such slender links, as a conformity with what some are pleased to call the essentials of religion. I was long fettered by these things, but now thank God they are placed upon the same shelf with the swaddling clothes which bound up my infancy." In the same year, Tyndall referred to Carlyle's *Chartism* as a "noble production," fully aware that to many it would appear "impious." "I however thank the gods," Tyndall exclaimed, "for having flung him as a beacon to guide me amid life's entanglements." Although Carlyle, as well as Emerson, Fichte, and other idealists helped give Tyndall some guidance, he believed that many years would pass before he could have enough knowledge to consolidate his religious position. On 26 June 1847 he wrote in his journal, "I have not yet digested my creed into any tangible form, nor should I wish to do so."[26]

Tyndall had lost his faith not in the existence of God, but only in the Christian dogma that obscured humanity's view of the divine. In a letter to Hirst dated 1855 he described his ambivalent reaction to two young men distributing Methodist tracts. On the one hand, they were "fanatics." But on the other, Tyndall detected "beneath the wildness of the enthusiast's eye the working of that spirit which keeps the world out of the mud." Tyndall believed that his scientific studies had honed his intellect to a razor-sharp edge, but "there was something at the heart of these methodist fanatics which I lacked, and which I longed for" (*LWJT*, 70). Tyndall still yearned for the clear direction supplied by a more definite creed.

In the cases of Huxley and Tyndall, their religious development did not involve an acute crisis of faith but rather a slow process of disillusionment with Christian doctrines. During the forties and fifties they were slowly working out a replacement for the faith of their childhood. However, Stephen and Clifford experienced explosive crises that quickly transformed their religious beliefs. Both lost their faith in Christianity during the sixties in the rationalistic atmosphere of Cambridge. Unlike Huxley and Tyndall, Stephen and Clifford were not from outside the university system, and they therefore did not find themselves confronted by the same social barriers.

Stephen was born in 1832 to a family of upper-middle-class evan-

Leslie Stephen
A Photograph of 1902

gelicals who could trace their roots to the influential Clapham Sect. His grandfather was intimate with William Wilberforce and Zachary Macaulay. His father, James Stephen, was permanent secretary in the civil service at the Colonial Office and a member of the middle-class intellectual aristocracy. The Stephen home was a Christian one wherein sincerity and moral rectitude were held in high esteem. In 1850 Stephen began his academic career at Cambridge. Five years later he was ordained deacon, and the following year he became a fellow and tutor of his college. Stephen enjoyed Cambridge life, involving himself in long distance walking, rowing, and mountaineering. According to his sister Caroline, Leslie had no doubts at this time.[27] But between 1859, when he took priest's orders, and 1862, when he no longer felt able to conduct chapel services, Stephen was in the throes of a religious crisis.[28] Although Stephen later characterized his dark night of the soul as an exhilarating process of liberation, it is more likely that he was devastated by his encounter with doubt. A fellow Cambridge student reported that Stephen had even considered suicide when the crisis reached its most critical moments.[29] In 1864 he left Cambridge. "I now believe," he stated in 1865, "in nothing, to put it shortly; but I do not the less believe in morality" (*LLLS*, 144).

A number of forces had been at work on Stephen's mind. He had been reading philosophy, and by 1860 he had devoured books by Mill, Comte, Kant, Hamilton, Hobbes, Locke, Berkeley, and Hume. In 1892 Stephen recalled the eroding effects of philosophers who "endeavoured to upset the deist by laying the foundation of Agnosticism, arbitrarily tagged to an orthodox conclusion." Hamilton was mentioned by name. In his autobiographical *Some Early Impressions* (1903) Stephen pointed to Buckle, Darwin, Spencer, *Essays and Reviews* and Colenso as the powerful dissolvents of belief during the early sixties.[30] Of all the agnostics, Stephen seems to have been the most influenced by biblical criticism during the period when the faith of his childhood was subjected to intense questioning.

About the same time that Stephen was encountering religious doubts, Clifford first arrived at Cambridge. Clifford was born at Exeter on 4 May 1845. His father was a justice of the peace and a well-known citizen of the town. Clifford entered Trinity College, Cambridge, in 1863, won second wrangler honors in the tripos of 1867, and was elected to a fellowship one year later. During the late sixties, when he was still at Cambridge, Clifford experienced a profound religious crisis. Before he took his degree and for some time after he was a High Churchman. His close friend Frederick Pollock reported "when or how Clifford first came to a clear perception that his position of quasi-scientific Catholicism was untenable I do not exactly know; but I know that the discovery cost him an intellectual and moral struggle." Clifford then became the center of a group of Cambridge men who were carried away by the infinite possibilities of Darwinian thought. By the time he left Cambridge in 1871 to become professor of applied mathematics at University College, London, he had already established a fine reputation among leading scientists. For a man who was just twenty-seven years of age his success had been, as Hutton put it, "meteoric."[31]

From the appearance of *The Limits of Religious Thought* (1858) until the late sixties, controversy raged over the epistemological questions raised by Mansel. It was during this period that Huxley and Tyndall were forging their alternative to Christianity and Clifford and Stephen were experiencing the pangs of religious doubt.

William Kingdon Clifford
A Drawing

The Spencer-Mill Battle

In 1865 an eminent unbeliever became embroiled in the controversy begun by Mansel, and his opposition to Spencer's *First Principles* made it clear that some freethinkers did not accept neo-Kantian epistemology, whether or not it was used to attack Christianity. John Stuart Mill's publication of *An Examination of Sir William Hamilton's Philosophy* offered unbelievers an alternative to Spencer's response to Mansel's Bampton Lectures.

Mill had been interested in Hamilton and Mansel for some time, and he followed the course of the Mansel controversy closely. He definitely read Spencer's agnostic encyclical in *First Principles.* Although Mill judged Spencer's program of intended works for the *System of Phi-*

losophy to be "rather too ambitious," he nevertheless became a subscriber and received sections of *First Principles* as they were issued. In 1863 he wrote to Spencer, "I cannot refrain from saying that your 'First Principles' appears to me a striking exposition of a consistent and imposing system of thought; of which though I dissent from much, I agree in more." However, nearly a year later Mill was less enthused about *First Principles.* Upon rereading the work he wrote to Alexander Bain that "on the whole I like it less than the first time. He is so good that he ought to be better."[32]

Ironically, it was Spencer's hysterical fear of being regarded as a disciple of Comte's which drew from Mill a comforting reply indirectly pointing out the gulf between the philosopher of evolution and the "saint of rationalism." "No Englishman who has read both you and Comte," Mill wrote in 1864, "can suppose that you have derived much from him. No thinker's conclusions bear more completely the marks of being arrived at by the progressive development of his own original conceptions; while, if there is any previous thinker to whom you owe much, it is evidently (as you, yourself say) Sir W. Hamilton."[33] At the time, Mill was working on what was to become *An Examination of Sir William Hamilton's Philosophy,* wherein he subjected the thinker to whom Spencer "owed much" to a devastating critique.

The influence of Mansel and the Kantian tradition had led Spencer away from Mill's empiricism and toward the formulation of his agnostic manifesto. Mill's utilitarianism owed its form of unbelief to English empiricism molded by the Enlightenment and had more in common with French movements of thought such as Positivism rather than German rationalism. Mill cannot be classed as an agnostic, although he is important for an understanding of the development of agnosticism.[34] The majority of the agnostics learned much from Mill and saw their creed as a logical extension of his doctrines. However, the key element in the agnostic perspective, the stress on the limits of knowledge based on an examination of the structure of the mind, was not a part of Mill's epistemology. Agnosticism represented a marriage of Mill's empiricism and Kantian modes of thought.

Mill believed that for the Victorian thinker, there were only two consistent epistemological viewpoints from which to choose and that each logically led to a specific answer to the problem of knowledge of God. The Benthamites' empiricist epistemology denied that knowledge of God or any transcendental entity was possible. On the other hand, Coleridge and his followers, in adhering to what Mill referred to as an "intuitionist" position, certified that human beings could perceive transcendental things, including God. For Mill the intuitionist theory of knowledge not only logically justified Christian theism, it also provided the rationale for conservatism in the political world. Conservatives could resist reform by viewing intuition as "the voice of Nature and of God, speaking with an authority higher than that of our reason" and then claiming that their doctrines were intuitive truths that reflected the natural or divine order underlying the social structure.[35]

Mill, then, did not see the struggle between intuitionists and empiricists merely as an academic issue; in his view it had profound consequences for social, political, moral, religious, and scientific principles as well. Like Spencer, he believed that the key to undermining orthodox Christianity and conservatism in political and social thought was to demolish the epistemology on which they rested. Whereas the *Logic* was written to combat German philosophers as well as English thinkers like Whewell who used intuitionism in order to defend conservative and Anglican institutions during the forties, the *Examination* represented Mill's sense that it was necessary to return to the attack on the same front in the sixties. This time Hamilton and Mansel were the targets instead of Whewell.

Mill's chief concern was to point out that Hamilton and Mansel's seeming adherence to the doctrine of the relativity of knowledge, an important aspect of empirical epistemology, was in fact deceptive, and that the true result of Hamilton's "philosophy of the conditioned" was to render aid to intuitionism. Mill maintained that Hamilton and Mansel, by asserting that belief or faith can go beyond knowledge, or that we can have faith in what we do not know, had vitiated their "philosophy of the conditioned." Through a concept of belief, Mill charged, they had smuggled an element of intuition into their philosophy of religion. Mill's attack on Mansel's attempt to have it both ways—to deny knowledge of God but affirm his existence—is equally applicable to Spencer's belief in an Unknowable.

Furthermore, Mill rejected the whole basis of Hamilton and Mansel's approach to the relativity of knowledge. According to Mill, Mansel and Hamilton had merely proved the unknowability of an impossible fiction, an inconceivable abstraction that had nothing to do with the notion of a real God possessing real attributes.[36] Every point Mill makes about the fallacious nature of Mansel's arguments on the unknowability of God due to the structure of the mind could have been applied to Spencer's reasoning in *First Principles*.

Spencer replied to Mill in the July issue of the *Fortnightly Review* for 1865.[37] Both were anxious to emphasize the superficial nature of their differences, Mill writing to Spencer on 12 August 1865 that "from the first I have wished to keep the peace with those whose belief in a substratum is simply the belief in an Unknowable."[38] The controversy between them, which never became heated, was allowed to end peacefully and quickly. Mill and Spencer agreed that it would not be wise to treat the English intellectual world to the spectacle of two liberal thinkers at each other's throats.

Mansel had no such reservations about engaging Mill in battle, and he adamantly defended Hamilton in *The Philosophy of the Conditioned* (1866). Critics agree that Mill's *Examination* demolished Hamilton and the Scottish School of philosophy.[39] So vociferous was Mill's attack on the Kantian tradition in England that one scholar places the end of agnosticism with the publication of the *Examination*, four years before Huxley coined the term.[40] However, the statement that Mill destroyed the credibility of Hamilton must be qualified in light of the agnostics' use of his arguments for the limits of knowledge.

The original agnostics did not agree with Mill that Mansel's arguments on the unknowability of God were ineffective. They sided with Spencer on this issue. Clifford, Stephen, Huxley, and Tyndall were avid readers of Herbert Spencer's work. There is proof that Stephen, Huxley, and Tyndall all read *First Principles*. Stephen recognized that Spencer had used Mansel's ideas to develop his epistemology, and that this was a valid move on Spencer's part:

> Mr. Herbert Spencer, the prophet of the Unknowable, the foremost representative of Agnosticism, professes in his programme to be carrying 'a step further the doctrine put into shape by Hamilton and Mansel.' Nobody, I suspect, would now deny, nobody except Dean Mansel himself, and the 'religious' newspapers, ever denied very seriously, that the 'further step' thus taken was the logical step. Opponents both from within and without the Church, Mr. Maurice and Mr. Mill, agreed that this affiliation was legitimate. (*AA*, 9)

Huxley read and criticized the proof of *First Principles* (*LLTHH* 1:228). Almost everything that Spencer wrote was sent to Tyndall prior to publication for comments and criticism, and Tyndall apparently read *First Principles*.[41]

The section entitled "The Unknowable" met with Tyndall's approval in his humorous letter of 1861 to Spencer: "Where the devil are you! I wanted to say to you that I have been reading your book, but your locality ranks in my mind with the ultimate basis and raw material of all consciousness—Both are at all events equally *unknown*. Well I have been reading the book and a tarnation good book." Tyndall joked that Spencer could use "the wholeness with which I go along with you" as an illustration of the absolute.[42] Years later Tyndall's enthusiasm for

"The Unknowable" had not waned. In 1873 he told Hirst that "I have been reading Spencer's First Principles: the first part [i.e., "The Unknowable"] is unspeakably good."[43]

Limits and Antinomies

We can see how Huxley, Clifford, and Stephen preferred Spencer's adaptation of Mansel's epistemology to Mill's attack on its defects in their use of the ideas of limits and antinomies. Clifford, Stephen, Huxley, and Tyndall were all familiar with Mansel's thought either directly through a reading of *The Limits of Religious Thought* or indirectly by way of contact with Spencer's *First Principles* or Mill's *Examination of Sir William Hamilton.* By opening the door for public discussion of the connection between epistemology and knowledge of God, and by feeding into the empiricist tradition, Mansel provided the agnostics with material for the formulation of their theory of knowledge. By encouraging the agnostics to turn to or, in some cases, return to a study of German philosophy, Mansel left his individual stamp on English agnosticism. It was Mansel who drew attention to the notion of the antinomies as a means of undermining the authority of reason in religious matters, and hence, helped to stimulate a resurgence of interest in Kant and Hamilton.

Stephen and Huxley were most directly influenced by Mansel's attack on rational theology in his *Limits of Religious Thought.* Like Mansel, they took only the negative aspect of Kant's *Critique of Pure Reason,* that is, those sections, like the passages on the antinomies, that undermined speculative reason's authority in transcendental matters. We have already encountered Huxley's whimsical reaction to Mansel in his use of a metaphor drawn from Hogarth's *Canvassing for Votes.* Huxley's reading of Mansel reinforced what he had previously learned from Hamilton. Although he had encountered Hamilton as a youth, he believed that he had somehow "laid hold of the pith of the matter, for, many years afterwards, when Dean Mansel's Bampton Lectures were published, it seemed to me I already knew all that this eminently agnostic thinker had to tell me" (*SCT,* 236). Mansel's Bampton Lectures created an intellectual atmosphere that provided Huxley with new opportunities to use the notion of the limits of knowledge to attack the false pretensions of Christian theologians.

Stephen was quite familiar with the whole Mansel controversy. He wrote the article on Mansel in *The Dictionary of National Biography* and included a discussion on the Mansel controversy in his *English Utilitarians.* In his essay "An Agnostic's Apology" he declared that the essence of Mansel's argument was based on the agnostic principle "that there are limits to the sphere of human intelligence." He termed Mansel's arguments in support of orthodoxy "an anachronism" and judged them to be "fatal to the decaying creed of pure Theism, and powerless against the growing creed of Agnosticism." Mansel's Bampton Lectures revealed the cause of the decline of theology, upon which the intellectual justification of "pure theism" was based. "The true reason for the decay of theology in the dogmatic stage is," Stephen stated, "that it tries to overleap the necessary limits of the human reason."[44]

Again and again Stephen ridiculed Christian theologians and metaphysicians for falling into contradictions and absurdity when they went beyond the limits of knowledge. The question of an afterlife he deemed insoluble owing to "the constitution of the human mind." Theorizing about a divine being transgressed "the limits of the human intellect."[45] Like Kant and Mansel, Stephen believed that the appearance of "antinomies" signaled the vain attempt to go beyond the limits of knowledge. Moralists, in seeking the essence of right and wrong, inevitably found themselves "at once in that region of perpetual antinomies, where controversy is everlasting, and opposite theories seem to be equally self-evident to different minds." Metaphysicians, in their attempt to reach a world outside all experience, only plunge us "into the transcendental region of antinomies and cobwebs of the brain."[46] One of Stephen's key strategies in his fight with Christian theologians was to echo Mansel's endeavor to undermine the role of reason in religious matters.[47]

Stephen believed that one of the results of Mansel's work was "some impulse to the speculation of the rising generation. Hamilton and Mansel did something, by their denunciations of German mysticism and ontology, to call attention to the doctrines attacked." In Stephen's case, it was just after *The Limits of Religious Thought* appeared that he began to read Kant (and other philosophers) seriously, and surely this was not a coincidence. Like Mansel, Stephen viewed those sections of Kant's work on practical reason to be inconsistent with his criticism of pure reason. "Kant himself admits," Stephen avowed, "that knowledge must be in some sense deduced from experience; and if he managed somehow to get into the transcendental world of 'things in themselves,' that, as most critics think, was his weakness, if not his inconsistency." Stephen displayed a somewhat similar opinion of Hamilton when, at the end of his copy of "Philosophy of the Unconditioned," he penciled in the comment "he had a very sound argument—only rather spoilt."[48]

But although Stephen had some reservations about the ultimate position Hamilton and Kant adopted, he was more than willing to latch onto the negative and destructive side of their work. Mansel had, in Stephen's words, "adopted from Hamilton the peculiar theory which was to enlist Kant in the service of the Church of England." Like Mill he despised the use to which Kant was put by Mansel, but he did not agree with Mill that the Kantian arguments were fallacious. Instead, Stephen set about drawing conclusions favorable to scientific naturalism from the premises that he and Mansel shared with Kant. One of Stephen's major reservations about Mill's work stemmed from Stephen's belief that Mill and his English contemporaries shared a general lack of knowledge of German thought. "How much better work might have been done by J. S. Mill if he had really read Kant! He might not have been converted," Stephen exclaimed, "but he would have been saved from maintaining in their crude form, doctrines which undoubtedly require modification."[49]

Tyndall and Clifford did not make direct use of Mansel in their work. But their writings reveal a concern for the issues raised by Mansel. Like Huxley, Tyndall had encountered the Kantian tradition before Mansel's Bampton Lectures appeared. In his journal he noted that on 16 January 1849 he was poring over "the transcendental dialectics of Kant" and by the sixteenth of the following month he had finished Kant's *Critique of Practical Reason.*[50] In April of that same year Tyndall explained to Hirst why he was interested in Kant. "Emerson," Tyndall wrote, "is strongly tinged sometimes with the philosophy of Kant, and without an acquaintance with the critical philosophy of the latter it is almost impossible to unravel some of his passages." Kant helped Tyndall in his reading of Emerson, Carlyle, and other idealists, by discussing how "knowledge of space and time is not a knowledge of objects *from without but only a knowledge of our mental organization.*" Kant also provided a link between spiritual awareness and science, for he united "Emerson's expression 'know thyself' and 'study nature' " into "one maxim." At this point Tyndall felt he had "once gone through the philosophy of Kant but only superficially," and he looked forward to a time when he could give Kant his undivided attention.[51] By October of 1849 he had again plunged into a study of Kant. "I have commenced reading Kant," Tyndall wrote in his journal, "and by degrees the light is increasing."[52] Later, Tyndall talked about Kant's "constructive imagination" in solving scientific problems in tones very similar to his own notion of "scientific imagination."[53]

Neither was Sir William Hamilton a stranger to Tyndall before 1858. In fact, Tyndall had been introduced to Hamilton at the British Association meeting at Glasgow in 1855. Later, in 1866, he referred to Hamilton's notion of an unknowable God in a letter published in *The Spectator,* which revealed at least an awareness of Hamilton's theological views.[54] Tyndall was a firm believer in the notion of the limits of religious thought. He declared that "behind, and above, and around all, the real mystery of this universe lies unsolved, and, as far as we are concerned, is incapable of solution." In his MS Note Books Tyndall recalled that one of his main preoccupations had been the search for the proper limits of knowledge. "As I grew older," Tyndall affirmed, "I endeavoured to make a clear distinction between what I *knew*, or might possibly know, and what I did not know and had no hope of knowing. I endeavoured, that is to say to mark out for myself what has been since called by a celebrated friend of mine 'Die Grenzen des Erkennens.' " He shared with Stephen and Huxley the belief that since God is a transcendental entity who dwells in the realm beyond the limits of knowledge, we could have no knowledge of him. He spoke of God as a being who one can "neither analyze nor comprehend."[55]

Pollock contended that, with the exception of Huxley, Clifford had "a much fuller appreciation of the merit and the necessity of Kant's work than most adherents of the English school of psychology." The failure to include Tyndall and Stephen as further exceptions does not lessen the force of Pollock's claim. Like Tyndall, Clifford was happy to use the notion of limits of knowledge, although reformulated in different language, to attack the pretensions of Christian theologians. Clifford outdid his agnostic colleagues in his impudent denunciations of orthodox theology. One verse of his humorous song "Poor Blind Worm" delighted the fellows of Cambridge University, who had gathered for a dinner party in 1870. "If you and God should disagree/On questions of theologee,/You're damned to all eternitee,/Poor blind worm!" Clifford resented the dogmatic claims of Christian theologians and insisted on the "unfathomable" quality of God.[56]

Stephen once said that "metaphysical arguments are apt to take the form of disputes about words." Metaphysical language, to Stephen, was meaningless and barren of any positive content. The production of meaningless terms by thinkers engaging in controversy signified that those participating in the dispute had gone beyond the limits of knowledge and were knocking their heads against an antinomy. Clifford agreed with Stephen:

> First, let us notice that all the words used to describe this immortality that is longed for are *negative* words: *im*-mortality, end-*less* life, *in*-finite existence. Endless life is an inconceivable thing, for an endless time would be necessary to form an idea of it. Now it is only by a stretch of language that we can be said to desire that which is inconceivable.

But Clifford did not use the notion of antinomy in the direct manner of Stephen or Huxley. In fact, Clifford denied that Kant's notion of antinomy was legitimate on the grounds that to maintain that a contradiction in human thought will eternally exist falsely assumes the universal validity of our previous experience and power of conception.[57] The antinomy of today, to Clifford, may be, as in the case of the antinomy of space, resolved tomorrow. Clifford seems to have benefited most from Mansel and the Kantian tradition in his development of an epistemology in connection with a philosophy of science.

Christian Contradictions

Mansel's attack on rational theology through the use of the strategy behind Kant's antinomies was also broadened by Stephen and Huxley into a means of turning the contradictory quality of Christian faith against itself. Stephen and Huxley were fond of extending a string of antinomy-like contradictions throughout an essay in order to drive the Christian reader into a corner, and they took malicious delight in fastening upon the sceptical arguments of fideists as support for their agnosticism.

Where Mansel had discussed contradictions in the theological notions of the absolute, the infinite, and the first cause, Huxley and Stephen pointed to the paradoxical quality of divine justice and the depiction of Christ in the Gospels in their main essays on agnosticism. Mansel had altered Kant's conception of the antinomies to begin with. Kant had not intended to give the antinomies an exclusively theological significance, as they represented for him the illusory quality that speculative reason had when it attempted to grasp the totality of conditions by conceiving of objects in appearances as things in themselves. But Mansel transformed the antinomies into paradoxical statements about God, and once the agnostics perceived the advantages of this type of argument for a proof of reason's impotence in religious matters, they were eager to apply it to other religious doctrines that they despised. In the agnostics' hands, the antinomies bore little resemblance to Kant's original notion.

Schurman perceptively points out that Huxley's analytic and iconoclastic genius led him to "revel in antinomies, and the method of his debate was to impale antagonists between the horns of an 'either-or.' "[58] In his essay "Agnosticism" (1889), Huxley attempted to undermine the doctrine of scriptural infallibility by setting up a series of contradictions inherent within the Bible itself. Huxley adopted Mansel's strategy of undermining reason to destroy the basis of Mansel's philosophy of religion. Using the story of the Gadarene swine (in Matt.

8:28–32, Mark 5:11–13, and Luke 8:30–33), Huxley discussed the belief in demons and evil spirits which pervades the Bible. Quoting from the Scriptures, he demonstrated that Jesus himself affirmed the existence of demons. This fact, to Huxley, generates a problematic situation. "I can discern no escape from this dilemma: either Jesus said what he is reported to have said," Huxley argued, "or he did not. In the former case, it is inevitable that his authority on matters connected with the 'unseen world' should be roughly shaken; in the latter, the blow falls upon the authority of the synoptic Gospels." The antinomy involves the following contradiction: if the New Testament reported correctly about Jesus, then it would follow that Jesus, the supposed God-man, believed in demons. If one wishes to save Christ from the depths of superstitious ignorance, it is at the cost of the authority of the Bible. "The choice then lies between discrediting those who compiled the Gospel biographies and disbelieving the Master," Huxley announced, "whom they, simple souls, thought to honour by preserving such traditions of the exercise of his authority over Satan's invisible world. This is the dilemma" (*SCT*, 218–20).

Huxley insidiously attempts to force the believing Christian to give up the Gospel writers rather than degrade their image of the perfect God-man. Once they do, Huxley pushes them on:

> After what has been said, I do not think that any sensible man, unless he happen to be angry, will accuse me of "contradicting the Lord and His Apostles" if I reiterate my total disbelief in the whole Gadarene story. But, if that story is discredited, all the other stories of demoniac possession fall under suspicion. And if the belief in demons and demoniac possession, which forms the sombre background of the whole picture of primitive Christianity, presented to us in the New Testament, is shaken, what is to be said, in any case, of the uncorroborated testimony of the Gospels with respect to "the unseen world"? (*SCT*, 228)

The contradictions lead one into the other until the Gadarene swine become a stumbling block to the doctrine of scriptural infallibility. Such cold-blooded dissection of Christian bibliolatry and the naive belief in miracles led Tyndall to refer to Huxley as "the arch-master of 'cut and thrust.' "[59] Evidently, Huxley enjoyed this type of argument, for he repeated it in his articles "Agnosticism and Christianity" (1889), "The Keepers of the Herd of Swine" (1890), and "Possibilities and Impossibilities" (1891) while he was embroiled in controversy with Gladstone (*SCT*, 193, 326, 371).

When Stephen's second wife died in 1895 he assuaged his grief by writing for his children the *Mausoleum Book*, an intimate and tender history of his marriages. "An Agnostic's Apology" was produced shortly after the death of his first wife, Thackeray's younger daughter, and it, too, was a sort of cathartic therapy for the heartbroken Stephen. Finding that the Christian doctrines of an afterlife and divine justice offered no consolation for the loss of his beloved, Stephen resolved to expose them as hollow mockeries. "An Agnostic's Apology" has a passionate and yet sarcastic tone that reflects the conditions under which it was written.

Stephen used Mansel's method of undermining reason's authority in religion by constructing antinomies centered on the problem of divine morality. It was Stephen's contention that unless we comprehend the problem of evil, then we must admit agnosticism. He began by asserting that if fate (or the will of God) rules the universe, then morality is impossible and we are not responsible for our actions. However, the accepted way out of this amoral position, the positing of free will, although relieving God of the responsibility for evil, does away with any notion of divine justice on earth. "The device justifies God," Stephen declared, "at the expense of making the universe a moral chaos" (*AA*, 22). Stephen argued that Christian theologians had recognized that reason is unable to grasp the problem of evil and is incapable of understanding how divine justice operates in this world. Their solution, according to Stephen, was to create a new subterfuge. After resolving the contradiction between fate and morality by creating the notion of free will, theologians were forced to overcome the resulting antinomy between free will and divine justice on earth by originating a new concept, that of heaven and hell:

> This world, once more, is a chaos, in which the most conspicuous fact is the absence of the Creator. Nay, it is so chaotic that, according to theologians, infinite rewards and penalties are required to square the account and redress the injustice here accumulated. What is this, so far as the natural reason is concerned, but the very superlative of Agnosticism? (27)

Stephen then turned to revelation in order to examine the possibility that here a more satisfactory solution is offered. But, he observed, to turn to revelation "is to admit that natural reason cannot help us; or, in other words, it directly produces more Agnosticism" (29). In addition, upon analyzing revelation, Stephen concluded that the notion of hell found in the Bible contradicted God's benevolence, and that to preserve God's goodness, one is forced into an inconsistent position. "Your revelation," Stephen declared, "which was to prove the benevolence of God, has proved only that God's benevolence may be consistent with the eternal and infinite misery of most of His creatures; you escape only by saying that it is also consistent with their not being eternally and infinitely miserable. That is, the revelation reveals nothing" (32). "An Agnostic's Apology" is one long question mark that leaves readers with nothing other than an awareness of their own impotence.

Stephen constructed a series of antinomies, as had Mansel in *The Limits of Religious Thought,* as part of a sustained effort to undermine systematically the authority of reason in the realm of religious matters. Both Huxley and Stephen attempted to demonstrate that there was no resting place between the principles of Christian theology and agnosticism. Once the reader was lured onto the horns of the first dilemma they were impelled logically toward agnosticism. Like Mansel, Huxley and Stephen offered a stark either-or, strict orthodoxy swallowed whole, contradictions, superstitions and all, or agnosticism. Stephen refused to consider the Broad Church as offering an authentic form of Christianity.[60]

Turning the Tables on Fideism

In addition to pointing to the paradoxes of Christian faith in order to undermine orthodox theology, Huxley and Stephen enjoyed drawing on the writings of Christian thinkers who dwelled on those contradictions so elusive for the weak reasoning faculty of human beings. The Christian fideist tradition made use of arguments other than Mansel's antinomies as a means to illustrate human intellectual impotence. But the agnostics' experience with Mansel led them to see how eminent fideists could become accomplices in the attack on the Church. J. H. Newman was particularly victimized by Huxley and Stephen.

Flint remarked in his *Agnosticism* that "the most ingenious and subtle arguments which have been urged against theism as a doctrine . . . have been oftener devised by theists than by anti-theists" (372). The point was made in the context of a discussion of the misguided philosophy of religious agnostics like Hamilton and Mansel, but it applies with validity to the whole fideist tradition. Stephen claimed that all Christian theology by its very nature contained a fideistic element. In his view, it was the orthodox Christians who were the sceptics because of their belief in miracles (which utterly destroyed the basis of science), their stress on humanity's need for revelation (which was to be accepted in spite of rational criticism), and their faith in an inscrutable being.[61] Stephen called into question the whole use of the label *sceptic* to describe unbelievers who supposedly doubted all, because the so-called sceptics of the day believed in the constancy or uniformity of nature, while their opponents expressed disbelief in the value of theories perceived to be dangerous to the Christian faith. The term *sceptic* should not be invoked in reference to the content of belief, rather, Stephen claimed it should be linked to the grounds for belief or disbelief. "If we insist upon using 'scepticism' to designate a mental vice, we must interpret it to mean, not doubt in general," Stephen affirmed, "but unreasonable doubt; and in this sense the most sceptical man is he who prefers the least weight of evidence to the greatest—or, in other words, he is identical with the most credulous" (*AA*, 46). Stephen had turned the tables on orthodoxy by redefining scepticism in such a way that Christian theologians were the great systematic doubters, and scientific naturalists the true believers.

"Is not the denunciation of reason," Stephen demanded, "a commonplace with the theologians? What could be easier than to form a catena of the most philosophical defenders of Christianity who have exhausted language in declaring the impotence of the unassisted intellect?" (*AA*, 7–8). When Stephen composed his list he included Butler, J. H. Newman, and Peter Browne, whose "attempt to out-infidel the infidel" was comparable to Mansel's.[62] In "An Agnostic's Apology" Stephen glibly pointed to Newman's lack of confidence in the capacity of unassisted reason to sufficiently support a belief in God, while in "Newman's Theory of Belief" (1877) he rated Newman as superior to Mill as regards their scepticism. "Here, as in so many cases," Stephen wrote, "the typical dogmatist is more sceptical than the typical sceptic" (*AA*, 10, 179–80).

Twelve years later Stephen looked upon Huxley's controversy with Wace over the Gadarene swine with great interest. On 8 April 1889 he wrote to Huxley and suggested to his friend that he look at some of Newman's works:

> I hope that you have made an end of the gentleman who believes in the pigs. It happens, however, that just after reading your article, I have come across a paper which falls in with it so oddly that I think it worth while to call your attention to it. In No. 85 of the Tracts for the Times (in Vol. V 1838–40) J. H. Newman wrote a very well written essay for the confusion of Protestants. Their argument was that the sacraments etc. were not provable by the New Testament. His reply is, no more are the doctrines of the trinity of Christ's divinity, or in particular, the admissibility of Gentiles to the church. His inference is, as there is no proof of either you may swallow both; but the curious thing is the clearness with which he shows that the gospels do not prove that Christ was more than a Jewish prophet of the usual kind—even taking them to be inspired. If you have to say anything more about it, it would perhaps be worth while to show that you have a cardinal to back you. (ICST-HP 27:57–58)

Huxley took Stephen's advice and, in "Agnosticism and Christianity" (1889), quoted "liberally" from Newman's works as support for his contention that the Bible, as a historical document, is no authority for accepting the reality of miracles. Newman was attempting to argue that faith should determine the true Christian's belief in miracles, but Huxley fastened upon Newman's attack on the inadequacy of external evidence, just as he used Mansel's undermining of reason in religious matters. In a footnote Huxley announced that "Tract 85 of the *Tracts for the Times* should be read with this *Essay* ['Essay on the Miracles recorded in the Ecclesiastical History of the Early Ages']. If I were called upon to compile a Primer of 'Infidelity,' I think I should save myself trouble by making a selection from these works, and from the *Essay on Development* by the same author" (*SCT*, 333).

The agnostics' strategy of using an eminent religious thinker with fideist sympathies as an accomplice proved to be effective, particularly in the case of Mansel and Newman. But Mansel was far more important than Newman for the development of agnosticism, because Stephen, Huxley, and Spencer placed his specific method for illustrating the weakness of human reason at the center of their theory of knowledge. Newman was referred to by Stephen and Huxley as a sceptic, but Mansel they claimed as one of their own, an agnostic.

Whether, as with Huxley and Stephen, the controversy surrounding Mansel's Bampton Lectures came after a long process of disillusionment with Victorian Christianity, or, as in the cases of Stephen and Clifford, the debate sparked by *The Limits of Religious Thought* coincided with an intense crisis of faith, Kantian concepts modified by Hamilton and Mansel played an important role in the development of agnostic thought. The agnostics did not share Mill's fear of borrowing from Christian thinkers. Encouraged by Spencer's successful use of Mansel in *First Principles*, Huxley, Stephen, Tyndall, and Clifford recognized the possible benefits to scientific naturalism in the construction of a new epistemology centered on the notion of limits of knowledge. But the fact that the agnostics could so easily place altered Christian ideas into the center of their thought may suggest that a significant religious dimension existed in agnostic theory. It is to the religious and theological elements in agnosticism that we now turn.

Chapter Five

RELIGION, THEOLOGY, AND THE CHURCH AGNOSTIC

> *Christianity was the last great religious synthesis. It is the one nearest to us. Nothing is more natural than that those who cannot rest content with intellectual analysis, while awaiting the advent of the Saint Paul of the humanitarian faith of the future, should gather up provisionally such fragmentary illustrations of this new faith as are to be found in the records of the old. Whatever form may be ultimately imposed on our vague religious aspirations by some prophet to come, who shall unite sublime depth of feeling and lofty purity of life with strong intellectual grasp and the gift of a noble eloquence, we may at least be sure of this, that it will stand as closely related to Christianity as Christianity stood closely related to the old Judaic dispensation.*
>
> JOHN MORLEY

In 1885 the *Agnostic Annual* reported that a movement was afoot to found an agnostic temple. "The first attempt at organisation on avowedly Agnostic principles is about to be made in the South of London," the journal announced, "where several gentlemen are endeavouring to establish what they purpose [sic] calling THE AGNOSTIC TEMPLE. The object of the organisation will be to disseminate a knowledge of the teachings of Agnosticism by the distribution of literature, the holding of meetings, etc."[1] The organizers took great care to stress the refined nature of their temple, not only by maintaining a discrete distance from lower-class religious radicals, but also by offering a cultured program for regular weekly meetings consisting of music, readings, and a short address. Agnosticism could be made respectable if it were patterned after the familiar forms of Christian institutions.

The founding of an agnostic temple was only one illustration of the religious dimension in Victorian agnosticism. Traces of religious and Christian elements also can be found in agnostic musings on the religion of the future, their views on authentic religious feelings, and their reaction to Spencer's worship of the Unknowable.

Scientific Naturalism, Social Context, and Intellectual Continuity

Frank Turner's assessment of the social significance of scientific naturalism would seem to discourage any effort by the historian to find in agnosticism important vestiges of religious or Christian beliefs. If the agnostics were committed middle-class scientific naturalists, then they, too, were caught up in the war against the Church as a means to undermine the intellectual authority of the old order. Any form of compromise by scientific naturalists yearning for the old faith would appear to be nothing short of traitorous.

But whereas the emphasis is on change in Turner's analysis of the shift of authority from one intelligentsia to another, Robert Young stresses the line of continuity running from natural theology to scientific naturalism. Despite Turner and Young's agreement that scientific and religious beliefs must be viewed in relation to the social context, they have presented two seemingly opposed interpretations of the ideological ramifications of scientific naturalism. Young's neo-Marxist approach is a fuller development of hints thrown out by Engels and Lenin. "Agnosticism," Engels sarcastically remarked, "though not yet [after 1851] considered 'the thing' quite as much as the Church of England, is yet very nearly on a par, as far as respectability goes, with Baptism, and decidedly ranks above the Salvation Army." Lenin was more explicit in his attack on agnosticism as a subsection of "empirio-criticism" which was merely another form of reactionary idealism. "Behind the epistemological scholasticism of empirio-criticism," Lenin declared, "one must not fail to see the struggle of parties in philosophy, a struggle which in the last analysis reflects the tendencies and ideology of the antagonistic classes in modern society."[2]

Agreeing with Lenin and Engels that an examination of the social implications of agnosticism reveals their conservatism, Young has argued that the scientific naturalists and Christian theologians were merely fighting over the "best ways of rationalizing the same set of assumptions about the existing order. An explicitly theological theodicy was being challenged by a secular one based on biological conceptions and the fundamental assumption of the uniformity of nature."[3] Although scientific naturalists rejected the usual theological justification for the status quo, they still attempted to reconcile people to the existing social order by conceiving of society as an organism that should be allowed to develop on its own accord since it is slowly progressing and growing due to the irresistible movement of natural laws.

A stress on continuity confronts historians with a number of striking images that invert our usual manner of perceiving key events in Victorian intellectual history. The spectacle provided by the meetings of the Metaphysical Society is not symbolic of the clash of science and religion. From the point of view of emphasizing continuity in Victorian thought, the Metaphysical Society is bourgeois society in miniature and represents squabbles from within the ruling classes on how best to rationalize bourgeois values. The famous debate between Huxley and Samuel Wilberforce during a British Association meeting at Oxford in 1860 does not encapsulate the conflict between evolution and Christianity. Rather, the significance of the debate lies in Huxley's appeal to the evangelical value of speaking truthfully in his response to Wilberforce. Halévy put forward the thesis in 1913 that, of all the countries of Europe in the first half of the nineteenth century, England was the freest from revolutions and violent crises because of the pervasive influence of the evangelical movement.[4] But Christian evangelicalism was not nearly so strong and vital by the mid-century despite repeated periods of revival and renewal. "Soapy" Sam Wilberforce was a pale evangelical imitation of his father, William, the great force behind the Clapham Sect. Yet England's unique stability lasted well into the century and beyond, and Young has offered us an intriguing explanation as to why. Science becomes the new evangelicalism and purveyor of the evangelical values of seriousness and duty which help England avoid violent upheaval during the latter half of the nineteenth century. The agnostics were nearly all raised as evangelicals, and they carried some of the attitudes of evangelicalism with them into their later lives. William Wilberforce's son passed on the mantle of evangelicalism to Huxley during their debate in 1860.

The relatively conservative quality of scientific naturalism can be illustrated by briefly comparing it to the ideology of a group of middle-class scientists in another European country. The idea of science that the German scientific materialists developed was shaped by the social and political turmoil of the 1830s and 1840s. The failure to achieve a unified Germany ruled by a popularly elected parliament left liberals like Vogt, Moleschott, and Büchner with little outlet for their ambition to participate in major political decisions, and their writings were an attack not only on religious but also political authority. Although they did not advocate the use of force in order to gain political advantage, the political consequences they drew from their scientific materialism were radical enough to deserve attention from the authorities. Vogt, a constant target of police harassment, fled Germany when the rump of the National Assembly, to which he had been elected, was forcibly dissolved by Prussian troops in 1849. He lived out his days as professor of geology and paleontology at Geneva. Political pressure led Moleschott to leave Germany for Zurich in 1856, and after years of wandering he found a new homeland in Italy, where he died in 1893.

Turner and Young have supplied us with two approaches to scientific naturalism, and therefore agnosticism, which seem, at first glance, to be contradictory. Where Turner emphasizes the notion that scientific naturalism was a substitute or replacement for conventional metaphysical beliefs based on Christian theology, Young looks for a basic continuity between the two ideologies. However, these two interpretive approaches need not be viewed as irreconcilable, and it is possible for historians to apply insights derived from both in order to enrich their understanding of what was a complex social, political, and intellectual process.[5] Young discusses the continuity of ideologies based on natural theology and scientific naturalism, but he does treat them as representative of two social orders and intellectual frameworks. Turner's interest in the sociology of intellectual change leads him to focus on how English society moved from one order to the next.[6] He therefore dwells on Huxley's role as an outsider during his radical, hungry youth, his fight against dogmatic Christianity, and his perception that scientific naturalists were offering a new leadership that would inculcate a modern set of values derived from the "new Nature." Young, on the other hand, centers on the more mature Huxley, no longer an outsider but now a member of the "establishment." As a body of doctrine designed to provide middle-class scientists with an air of authority, scientific naturalism proved to be extremely successful. This is the Huxley who was a fellow of Eton, who received a Civil List pension, who was consulted by Lord Salisbury, then Conservative premier, about scientific policy and appointments, and whose lean and hungry look had become replaced by a stoutness tending toward corpulence. This Huxley no longer shocked the Victorian public, and having won his battles he could admit to his strong religious nature.

Pope Huxley and Original Christianity

Although the agnostics borrowed from Mansel, the Kantian tradition, and fideism in order to attack the authority of the Church, the stress on the limits of human knowledge and God's corresponding unknowableness represented only one element of Christian thought which the agnostics retained in their views on religion, ethics, and science. The agnostics all came from Christian households, and they shared many of the values espoused by Victorian Christians. Many of the agnostics revered the Bible as a reservoir of spiritual truth. Although they experienced a moral revulsion to those Christian doctrines most readily identified with evangelicalism, they still retained an evangelical fervor for sincerity, honesty, and moral earnestness. The agnostics all lived model lives of respectability. Frederick Pollock once remarked that there was "enough goodness in Huxley to make all England Christian, if it could only be parcelled out and distributed around."[7] However, it is easy to dismiss these vestiges of Christianity in the agnostic mentality as being of little consequence. Christian doctrine, Blyton argues, was thrown overboard by the agnostics, and a secularized version of Christian ethics was retained. The agnostic use of biblical language and ideas can be interpreted as a purely polemical strategy since prose shaped by biblical style and rich with allusions drawn from Holy Scripture had a powerful effect on the Victorian public.[8]

One author has dubbed Spencer, Darwin, Tyndall, and Huxley as the "four Evangelists" of agnosticism.[9] In a similar vein Clifford has been called "an apostle of scientific thought," Stephen a "Hebrew prophet" and Tyndall the "Apostle of Physical Science."[10] Huxley's missionary spirit has also been noticed. He has been described as the "great apostle of the modern gospel of science," "the John Knox of Agnosticism," "prelate," "priest," and "prophet of science," whose lay sermons presented a "Creed of Science for its Thirty-nine Articles."[11] But such satirical references to the agnostics are more often than not meant to indicate that their manner or method of disseminating the good news of modern science resembled the Christian preaching of the gospel. There is rarely a serious intent to imply that the agnostics preserved any substantial religious content from Christianity.

In the heat of controversy Huxley tended to preach his message dogmatically, which led Hutton to playfully name him "Pope Huxley." Huxley, Hutton maintained, responded to criticism "in the tone of a Papal bull,—containing violent censures . . . as well as dogmatic decrees."[12] Hutton chided Huxley for being untrue to the agnostic attitude of suspended judgment in the face of lack of evidence. Although Hutton joked about Huxley's affinities with the very theologians he attacked, he did not allow his sense of irony to obscure the important debt the agnostics owed to Christianity and the vital religious content of their thought. When Huxley talked of his deep religious sensibility in his essays, Hutton, rather than claim that the agnostic was merely offering a sop to public opinion, took Huxley's religiousness seriously. Hutton sensed an ambivalence in Huxley toward religion. "In our belief," Hutton declared, "Professor Huxley had a half-unconscious craving, to which he thought it wrong to give way, for that passionate faith which he said that he desired to undermine in all cases in which there was, in his opinion, no possibility of what he termed verification. Indeed, his heart often rose up in insurrection against his scientific genius, and compelled him to feel what was entirely inconsistent with the logic of his thoughts."[13]

Another contemporary of Huxley's who agreed with Hutton's estimate was Wilfrid Ward (1856–1916), a Catholic and later editor of the *Dublin Review*. Ward was Huxley's neighbor at Eastbourne. Throughout the last years of Huxley's life, the two men had intimate talks devoted entirely to religious issues, and during these, Ward was surprised to learn of the agnostic's devout nature.[14] Ward interpreted Huxley as one who was torn between the destructive quality of the theoretical conclusions he drew from his agnosticism and his practical attitudes that drove him toward theism. "I concur with those who believe that his rooted faith in ethical ideals," Ward asserted, "which he confessed himself unable to account for by the known laws of evolution, implied a latent recognition of the claims of religious mystery as more imperative and important than he could explicitly admit on his own agnostic principles. Careful students of his writings are aware how far more he left standing of Christian faith, even in his explicit theories, than was popularly supposed; and this knowledge appeared more and not less significant to some of those who conversed with him on these questions."[15] It was Ward's opinion that Huxley's combativeness was a result of the bitterness he still felt when he remembered the intolerance he experienced as a youthful scientist fighting every inch of the way against bigoted theologians.[16] In those days, Huxley once told a friend, "men like Lyell and Murchison were not considered fit to lick the dust off the boots of a curate. I should like to get my heel into their mouths and scr-r-unch it round."[17] But beneath the smoldering animosity toward ecclesiasticism in Huxley's later years, Ward could still detect an intensely religious soul. In many ways Huxley was representative of the majority of the agnostics.

The agnostics revealed their religious nature even when they were attacking the Christianity of their time. The basis of their criticism of the rigid dogmatism of the churches was their belief that Victorian Christianity was a perversion of the original, pure religion as founded by Christ. "The Church founded by Jesus," Huxley wrote in a letter of 1889, "has *not* made its way; has *not* permeated the world—but *did* become extinct in the country of its birth." Huxley, like many of the other agnostics, genuinely revered Christ and his teachings. According to Hutton, Huxley frequently indulged in sudden bursts of passionate feeling for Christ. Tyndall also saw in Christ an attractive symbol of true religious ideals in contrast to the degenerate state of present day Christianity. He wrote in 1848 that "the Great Spirit which from time to time expresses himself audibly among the sons of men dwells far below the scum of sects—into this shall methodism, churchism and many other isms one day sink and a purer lovelier and more practical faith—a faith which Jesus taught and John understood shall bend with benignant influence over our altered world." Stephen put forward a similar idea. The ancient creeds, he asserted, "were indeed in great part the work of the best and ablest of our forefathers; they therefore provide some expression for the highest emotions of which our nature is capable."[18]

Even Clifford, who frequently indulged in savage attacks on Christian orthodoxy which outdid all of the other agnostics in their ferocity, shared with Tyndall, Huxley, and Stephen a high regard for the original spirit of Christianity. When Clifford was wasting away in 1878 of the consumption that would claim his life a year later, he could still summon the strength and wit to answer to a newspaper report that he was converting back to Christianity. Flatly contradicting the story he stated that his "M.D. had certified he was ill, but 'twas not mental derangement."[19] Clifford saw Christianity as an idolatrous religion, barely distinguishable from the pagan abominations condemned in the Bible (*LE* 1:252). In 1869 he satirized ceremonies to install a new bishop. "The entire town is in an uproar for the ecclesiastical fuss that is to take two hours in the streets and the cathedral tomorrow," Clifford wrote, "enthronization of the new bishop, parade through the public ways of him and minor fetishes, as the mace, cocked hat, and Sword of the Civic Functionary, and subsequent grand banquet to the priests of Baal."[20] Clifford likened present-day Christian practice to forms of worship found in ancient Egypt. The Church developed a creed very different in substance from Christ's message, and in his essay "The Ethics of Religion" Clifford had only praise for the Sermon on the Mount. "The gospel indeed came out of Judaea," Clifford affirmed, "but the Church and her dogmas came out of Egypt" (*LE* 2:230).

The fount of the original spirit of Christianity, the Bible, was regarded by the agnostics as a book of great wisdom and beauty. In his attacks on the doctrine of scriptural infallibility Huxley was combating what he conceived to be a tendency to erect the Bible into an idol that destroyed the deep richness he earnestly desired to preserve. He praised the Bible for its simple honesty, its "moral beauty and grandeur," and even its scientific methodology. "As to the methods by which the Biblical writers arrived at their great truths," he wrote, "I do believe that they were in the truest and highest sense scientific. I recognize in their truths the results of a long and loving, if sorrowful, study of man's nature and relations." But although the Bible followed sound inductive principles, it never claimed for itself scientific authority.[21] Tyndall, who knew the Bible by heart as a boy, also revered the Bible, and considered "the purity of the Scriptures one of the highest proofs of their divine origin."[22]

The agnostics agreed that the Bible, if read without prejudice, was the best antidote to bibliolatry. Huxley argued that "the Bible contains within itself the refutation of nine-tenths of the mixture of sophistical metaphysics and old-world superstition which has been piled around it by the so-called Christians of later times" (*SCT,* 268). Huxley distressed some of his freethinking friends during his tenure as member of the London School Board in the early seventies, for he sided with those who believed that the Bible might be read in the public schools. But when Stephen heard of it he responded, "What made us freethinkers? Why, reading the Bible!"[23] The Bible itself was a revolutionary book.

The agnostics did not believe that the answer to the disintegration of original Christianity was to work within the system and preserve it through modernization. The liberal Christian attempt to update the Church was doomed to failure because the Christian religion had sunk too low to be revived. The distortions of Christ's teachings had, through the ages, become too ingrained in the heart of the Church to be removed piecemeal. Stephen argued that the belief in hell was part of the "very structure of Christianity" and therefore could not be arbitrarily excised. "The whole must require to be remodelled," Stephen insisted. "We cannot retain the amiable parts of a doctrine whilst leaving out the sterner elements, or be sure that we can clip and mangle without emasculating" (*AA*, 105). Whether we read carefully Tyndall's attack on the supposed efficacy of prayer, or Stephen's repudiation of the notion of hell, or Clifford's disgust with the doctrines of original sin and vicarious sacrifice, the same theme emerges: the agnostic belief that Christianity had become vulgar, immoral, and hopelessly foreign to the true religious spirit of pure Christianity.[24]

The New Reformation

Attacks on dogmatic Christianity gave the writings of the agnostics an unavoidably destructive quality. However, although the agnostics viewed their critical efforts as necessary, they did not fancy themselves to be mere nihilists. They saw their negative comments as a needed preliminary to a positive attempt to construct a new religion. Tyndall wrote to Spottiswoode in 1877 that "my desire has been to act the part of a conservative rather than that of a Destructive, by gradually preparing the public mind for inevitable changes which without this preparation might take a revolutionary form." Wilfrid Ward noticed that Huxley "resented being identified with simple destruction in matters of religious faith." In 1892 Huxley referred to the aphorism by Cuvier, prefixed to the prologue of *Controverted Questions*, which stated "one should clear the ground before beginning to build." Huxley affirmed that this aphorism represented both the positive and negative quality of his purpose during the last thirty years. "It will be observed," Huxley went on, "that it enjoins the clearing of the ground, not in a spirit of wanton mischief, not for destruction's sake, but with the distinct purpose of fitting the site for those constructive operations which must be the ultimate object of every rational man. Neither one lifetime, nor two, nor half a dozen, will suffice to clear away the astonishing tangle of inherited mythology."[25]

As constructors of a new religion, the agnostics, in particular Tyndall and Huxley, perceived themselves to be religious reformers like Luther. Huxley believed that the revolution effected in the modern mind by the beneficial impact of science represented the final climax of the Protestant Reformation. "The act which commenced with the Protestant Reformation is nearly played out, and a wider and deeper change than that effected three centuries ago . . . is waiting to come on." Just as Luther had contrasted free thought to traditional authority in order to undermine the strength of the decaying Catholic Church, Huxley saw a new movement at work which insisted "on reopening all questions and asking all institutions, however venerable, by what right they exist, and whether they are, or are not, in harmony with the real or supposed wants of mankind" (*SE*, 191–92). In the preface to *Science and Christian Tradition* Huxley referred to this revolution, which embodied the spirit of intellectual freedom born of science, as the "New Reformation" (vi).

Huxley and Tyndall were quite fond of drawing a comparison between their efforts to bring about a new reformation and the beginnings of Protestantism. In 1849, taking advantage of some free time while he studied at Marburg, Tyndall went, as he put it, on "a pilgrimage to the scenes of Luther's life." He reported to Hirst that he went to Eisenbach to see the room where "Luther flung the inkbottle at the devil," and then traveled to Wittenberg to see Luther's grave, his house, and old furniture.[26] In 1874, when Tyndall was at the height of his career and was about to deliver his notorious "Belfast Address" as the incoming president of the British Association for the Advancement of Science, he drew a parallel between the opposition he anticipated from bigoted Christians and that which Luther encountered from dogmatic Catholics. "I will go to Belfast as Luther did to Worms if necessary," Tyndall wrote, "and meet if requisite all the Devils in Hell there."[27] In support of his attack on the rigid emphasis by Victorian Christians on custom and ritual, in particular in the case of the observance of the sabbath,

Tyndall quoted from Luther and Melanchthon to demonstrate that "the early reformers emphatically asserted the freedom of Christians from Sabbatical bonds" (*NF*, 16).

Huxley also felt a sympathetic bond between himself and Luther. In 1847, while on his *Rattlesnake* voyage, Huxley discussed his religious doubts and difficulties as if they were, like Luther's position at Worms on the sins of Catholicism, the result of honest and sincere reflection. Huxley wrote in his journal, "Ich kann nicht anders! Gott hilfe mir!'" But later, in "Agnosticism: A Rejoinder" (1889) Huxley found in Luther a vindication for his agnostic proclivities as well. He claimed that he had reached his agnostic standpoint through the exercise of his private judgment. "My position is really no more than that of an expositor," Huxley declared, "and my justification for undertaking it is simply that conviction of the supremacy of private judgement . . . which is the foundation of the Protestant Reformation."[28]

The new reformation could involve the founding of new, pure institutions to replace the corrupt churches of the day. Huxley was quite serious in 1871 when he talked of the possibility of "the existence of an Established Church which should be a blessing to the community. A Church in which, week by week, services should be devoted, not to the iteration of abstract propositions in theology, but to the setting before men's minds of an ideal of true, just, and pure living. . . . Depend upon it, if such a Church existed, no one would seek to disestablish it."[29] Stephen was attracted to the founding of a Church which would be based on aesthetic principles, and he claimed that such an institution was the only hope for those who desired a truly catholic religion. He affirmed that "a dogma is only offensive when you are asked to believe it; but we may be all members of a Church in which a dogma is no more essential than a vestment, and is simply an arbitrary sign of certain emotions. Indeed, by this method we may reach a catholicism wider than has ever yet dawned upon the imagination of mankind" (*FP*, 56).

The Religion of the Future

The agnostics were not trying to destroy all forms of religion when they launched their onslaught on Christianity. The new reformation represented for Huxley the building of a new religion that would recover what had been lost by Christianity when it perverted the pure ideals of its founder. The agnostic faith in the continuing validity of some type of religion can also be perceived in their musings on the future of religion. An aphorism scribbled by Huxley in 1894 states that "the religion which will endure is such a day dream as may still be dreamed in the noon tide glare of science." Likewise, Tyndall talked confidently of the survival of religion. Since religion was "ingrained in the nature of man" it would be reconstructed, as it has been many times in the past. But Tyndall could not foresee the precise form it would take.[30]

Of all the agnostics it was Stephen who devoted the most energy to the subject of the religion of the future. The religious instincts of human beings, Stephen believed, were indestructible, and therefore they would persist if the Anglican Church, or even Christianity perished (*FP*, 7). But the nineteenth century was an age of change, caught awkwardly between lingering, ancient forms of faith, and the yet to be born religion of the future. "The old creed," Stephen stated, "elaborated by many generations, and consecrated to our imaginations by a vast wealth of associations, is adapted in a thousand ways to the wants of its believers. The new creed—whatever may be its ultimate form—has not been thus formulated and hallowed to our minds" (*FP*, 359). Stephen saw the problem as being manifested in a painful discord between the imagination, which was essentially conservative in nature due to its attraction to the old symbols and dreams, and reason, which was progressive in its construction of a new order with the aid of science. The new order did not appeal to the imagination but remained "colourless and uninteresting, because the old associations have not yet gathered round it." Its only resort was an "appeal to its utilitarian triumphs in order to gain allies against the ancient idolatry" (*HETEC* 1:14–17).

In his endeavor to understand the difficulty of constructing a new faith Stephen turned to a study of the past. Stephen's *History of English Thought in the Eighteenth Century* had two objectives. First, by rescuing the old deists from oblivion, Stephen hoped to demonstrate that orthodoxy already had been bested intellectually by eighteenth-century unbelievers. Second, Stephen wished to comprehend why evangelicalism, rather than the superior position of rationalists, had captured the minds and imaginations of the bulk of the English people in the last century. Stephen attempted to uncover the cause of the failure of eighteenth-century rationalism in England in the hope of avoiding a similar failure in his own day.[31]

Stephen also learned what to expect in the future from his study of the development of Christianity. Here was an intellectual movement that had succeeded where the deists had failed. Since he was dealing with a far longer stretch of time in this case in comparison to his work on the eighteenth century, Stephen found more scope for the application of evolutionary theory. He viewed the entire history of religion as subject to the process of natural selection. Christianity had won the struggle for existence among religions not because of some claim to perfect truth, but because it suited the conditions of the time and the needs of the society in which it originated.[32] "We can only explain the spread of the organism by showing how and why the soil was congenial," Stephen exclaimed. "The Christian doctrine obviously spread, as every doctrine spreads, just so far as it was adapted to men at a given stage" (*AA*, 312). Christianity was successful because countless multitudes found in it what they wanted, not because of the personal characteristics of its founders. It was not only appealing to the imagination of the populace, but it was vague and open-ended enough to attract the higher intellects, who erected an elaborate theological structure that satisified their reason. Just as explicable on the basis of evolutionary theory was the decline of Christianity. A result of a particular set of social conditions, Christianity's remedy was no longer appropriate to modern needs. The excessive transcendental element of Christian thought, which in the beginning gave free play to popular imagination and the reason of the intellects of the time, later became a source of decay.[33]

Stephen believed that with the decline of Christianity and the corresponding vacuum that now existed, the situation resembled the period preceding the birth of Christianity (*AA*, 353). But the question still remained, who would win the struggle for existence today as the Christians had centuries ago? Stephen asked, "What sect is analogous to the ancient Christians? Who are the Christians of the present day? Which, in all the huddle of conflicting creeds, *is the one* which is destined to emerge in triumph?" (*AA*, 354–55). Stephen did not profess to know the answer. He believed that "the problem about the religion of the future is simply insoluble. Inspired prophecy is out of date" (*AA*, 342). However, from his examination of the evolution of religion, the origins of Christianity, and eighteenth-century deism, he could conclude that the new religion would be a higher one in that it would satisfy modern reason and imagination in a more sophisticated manner than had Christianity (*AA*, 301).

Stephen saw his role in this evolutionary process as a modest one. At most the agnostic could contribute to laying a philosophical basis of a new religion. It would be premature, in his opinion, to propagate a fully developed religion. "We are only laying the foundations of the temple," Stephen declared, "and know not what will be the glories of the completed edifice" (*FP*, 360). But Stephen wanted to help shape the process that was giving birth to the religion of the future. He warned that agnostics must not only attack Christian dogmatism, but also become aware of the power of the Church's appeal to the imagination, which is achieved through an elevating morality, aesthetically pleasant forms of worship, and the attraction of a strong social bond. Stephen believed that it was important to "recognize what is good in the feelings to which the Church owes its strength, and show how they may be combined with full acceptance of the teachings of reason." The agnostic, if he wished to succeed in building a new religion, was charged with the duty of speaking the truth in order to provide "new channels for the utterance of our emotions." Plainspeaking was the path to a new set of symbols attractive to the imagination and yet based on the advances of reason. Stephen affirmed that "the more we really believe that religion is founded upon enduring instincts which will find an expression in one form or another, the less anxious we should be to retain the old formulae, and the more confident that by saying what we think, in the plainest possible language, we shall be really taking the shortest road to discovering the new doctrines which will satisfy at once our reason and our imagination."[34]

The Religion of Agnosticism

Huxley vehemently denied that agnosticism could be described as a "creed" and claimed that agnosticism had nothing to do with religion (*SCT*, 249–50, 310). However, Huxley, Tyndall, Stephen, and Clifford all put forward their views on the nature of true religion, and they saw them as consistent with their agnostic principles. Their work contained a positive, constructive dimension that is often buried under their explosive attacks on Christian orthodoxy. Up to this point we have discussed their call for a new religion and their vision of a religion of the future. Now we must turn to the actual content of their religious thought.

Stephen, Clifford, Huxley, and Tyndall all believed that religion fell within the realm of feeling, emotion, imagination, inwardness, and symbol. In acting as an outlet for our inner feelings, religion was akin to art and poetry. The religious instinct, inherent in human beings, was a living, growing thing both within the individual and in the race.

Huxley's lifelong interest in religion is illustrated by his preoccupation with religious themes in his published works as well as in his unrealized plans for a comprehensive history of Christianity based on a long and serious study of theology.[35] This fascination with religion was the result of Huxley's own strong religious sensibility. In 1873 he answered Galton's request for information on his character by noting his "profound religious tendency capable of fanaticism, but tempered by no less profound theological scepticism."[36] The source of Huxley's religiousness was an awareness of an impenetrable mystery that evoked a religious response in all of humanity. Hovering over the "abyss of the unknown and unknowable," human beings were but dimly graced with insight into the world. "But in this sadness," Huxley believed, "this consciousness of the limitation of man, this sense of an open secret which he cannot penetrate, lies the essence of all religion; and the attempt to embody it in the forms furnished by the intellect is the origin of the higher theologies" (*MR*, 33).

In addition to seeing religion as a proper manifestation of that "sense of an open secret," Huxley emphasized the moral dimension of religiosity. Religion involved "the reverence and love for the ethical ideal, and the desire to realise that ideal in life." Huxley's vehement statement that "religious feeling" is "the essential basis of conduct" is more readily explicable in light of his view of the link between religion and morality.[37]

Huxley never systematically outlined the content of his religious feelings. These matters were better left unsaid unless the agnostic was to become as dogmatic as the orthodox. However, Huxley hinted at his general position. In a letter to Romanes in 1892 he wrote: "I have a great respect for the Nazarenism of Jesus—very little for later 'Christianity.' But the only religion that appeals to me is prophetic Judaism. Add to it something from the best Stoics and something from Spinoza and something from Goethe, and there is a religion for men" (*LLTHH* 2:361).

In his eagerness to leave no place for religion in agnosticism, Benn pointed to Leslie Stephen as the best example of a consistent antireligious agnostic.[38] However, Stephen repeatedly affirmed his belief in the value of religion. Of all the agnostics Stephen was the most careful to identify religion almost entirely with ethics and aesthetics.[39] Religion's role was to express moral sentiments in a beauteous fashion and thereby to inspire the individual to greater ethical heights. Religion, to Stephen, must become "the embodiment in concrete images of the spiritual aspirations of mankind" and must be built upon "a purely aesthetic basis" (*FP,* 55–56). Stephen believed that feelings, especially religious ones, were best voiced through the medium of poetry.

In Stephen's view Christianity was poetic in substance. But instead of treating religion as symbolic of elevated feelings and ideals, Christianity tended to equate religious doctrine with literal truth. As a result, at the heart of orthodox Christianity there lay an unpoetic materialism that offered homage to grossly material symbols; orthodox Christianity crudely interpreted the articles of its creed in physical terms. Compared to the supposedly materialistic scientists, Christianity was far worse. There is more materialism, Stephen declared, "in popular sentimentalisms about the 'blood of Jesus' than in all the writings of the profane men of science." Stephen's advice to Christianity was to admit that its truth was poetic or symbolic, not literal or scientific. If the Church were willing to treat its truths as beautiful legends and imaginative ideas, then a reconciliation between faith and reason could be found which would protect the divine from the sceptic, because the demand for proof would no longer be relevant.[40]

Clifford also could be surprisingly eloquent on those rare occasions when he addressed the topic of authentic religion. Pointing to Maurice, Kingsley, and Martineau as examples of Christians who put forward "forms of religious emotion which do not thus undermine the conscience," Clifford retained a place for religion in his agnosticism. He referred to religious feeling as "cosmic emotion," by which he meant a sense of awe in regard to the order manifested throughout the universe (*LE* 2:242, 253–85).

Tyndall was enthralled by the religious aspect of idealism which he found in Carlyle, Emerson, Fichte, and German philosophy. As a result, Tyndall constantly stressed the need for a vital commitment of one's whole being to religion, and he pointed out the error of identifying religion with intellectual persuasion. Tyndall declared that "religion is not a *persuasion*, it is a *life*," that "*it* must come from the heart," and that religion "finds a root in human nature which is deeper than all sensuous experience and lies below our modern science of logic."[41]

To Tyndall it was this personal religious experience in the individual which was significant and not the attempt to fossilize heartfelt religion into specific forms. During a discussion with Hirst at Queenwood in 1852 Tyndall used the image of stick and vine to represent his view on the relationship between form and religion. "Forms bear the same relation to vital religion as the stick does to the vine which it supports—the life is in the vine, but had it not the support to cling to it might grovel on the ground and its fruit be spoilt. But just as there are vines which spring up by the rock itself and cling thereto needing no stick, so there are men whose religion needs no stereotyped form." Five years later Tyndall wrote in his journal that he surprised two pious Americans "for I could talk to them of religious experience which went nearly to the bottom of theirs, and at the same time seemed to regard their forms, which they deemed essential, very lightly. Indeed what I would call a form, for example Christianity itself, they deemed the essential marrow of the thing."[42] Religion could not be frozen into a specific form because, by its very nature, it was a living, growing spirit. Religious emotion, ordinarily a valid form of human experience, became a falsity when formalized into statements about objective reality.

Tyndall found that many of his contemporaries were unwilling to accept his view of religion. In 1879 he referred to "the religion which Mr. Huxley and myself favour" and attacked those like Mivart who were so narrow-minded that they were unable to understand either his religion or "the religion of a Fichte, of an Emerson, or of a Carlyle."[43] In order to defend the validity of his conception of religion Tyndall compared the "rigid" religious symbols "of our present christianity" to Christ's attempt to "infuse a spirit into a world which has lost its way in a labyrinth of Formulas. He contended for an inner, vital, warming principle, which should be the spring of action to humanity."[44] Tyndall pictured Christ as one who broke the Hebrew sabbath deliberately as a defiant gesture against those who suffocated religion under a load of formulas, forms, and ceremonies (*NF*, 11–12).

Tyndall believed that the appropriate mode for expressing religious sentiment was poetry (*FS* 2:196). He foresaw an important role for the poet as the future bearer of religious culture to the world:

> To him it is given for a long time to come to fill those shores which the recession of the theologic tide has left exposed. Void of offence to science, he may freely deal with conceptions which science shuns, and become the illustrator and interpreter of that Power which as 'Jehovah, Jove, or Lord,' has hitherto filled and strengthened the human heart. (*FS* 2:99)

Turning once again to the original Christians in order to make his point, Tyndall argued that both Saint John and Christ used poetry to articulate religious truth (*FS* 2:357).

Theology, Science, and Religion

Whereas religion, along with poetry and art, belonged to the province of feeling, the agnostics placed theology in the realm of intellect or reason. The propositions of theology could therefore be tested like any other proposition in the realm of intellect. Theology must submit itself to the authority of science. Furthermore, on the basis of the distinction between religion and theology, the agnostics could claim that there was only a potential conflict between theology and science, not religion and science.

For Huxley, religion belonged to the realm of feeling, while science was part of the realm of intellect. Science and religion, if rightly conceived, could never come into conflict because each realm was distinct and without authority outside its proper sphere of interest. Huxley affirmed in 1859 that "true science and true religion are twin-sisters, and the separation of either from the other is sure to provide the death of both. Science prospers exactly in proportion as it is religious; and religion flourishes in exact proportion to the scientific depth and firmness of its basis."[45] However, theology, distinct from religion and operating in the world of intellect because of its claim to embody feelings in concrete facts, was potentially in conflict with science. Science and religion were only at odds if religion was wrongly identified with theology.[46] The real war was between agnosticism and ecclesiasticism ("the championship of a foregone conclusion as to the truth of a particular form of Theology"), and, parallel to this antagonism, between science and theology (*SCT,* 312). Huxley found the distinction between religious feeling and theological dogma in Carlyle and in German thinkers such as Goethe, to whom his reading of Carlyle had led him. "*Sartor Resartus* led me to know," Huxley wrote privately to Kingsley in 1860, "that a deep sense of religion was compatible with the entire absence of theology" (*LLTHH* 1:237).

Huxley first put forward his view of the distinction between religion and theology in an anonymous editorial in the *Reader* in 1864 titled "Science and 'Church Policy.'" "Religion has her unshakeable throne in those deeps of man's nature which lie around and below the intellect, but not in it. But Theology," Huxley declared, "is a simple branch of Science or it is nought; and the 'Church Policy' which sets it up against Science is about as reasonable as would be the advocacy of the claims of the rule of three to superior authority over arithmetic in general."[47] Twenty-one years later he was still claiming that there was no conflict between science and religion. "The antagonism between science and religion," Huxley announced, "about which we hear so much, appears to me to be purely factitious—fabricated, on the one hand, by short-sighted religious people who confound a certain branch of science, theology, with religion; and, on the other, by equally short-sighted scientific people who forget that science takes for its province only that which is susceptible of clear intellectual comprehension; and that, outside the boundaries of that province, they must be content with imagination, with hope, and with ignorance" (*SHT,* 160–61).

Stephen agreed with Huxley that the "conflict [was] between science and theology" (*FP,* 58). He, too, separated the emotional sources of a valid religion, and the questionable legitimacy of Christian theology. Tyndall presented the crucial difference between religion and theology by using a number of vivid contrasts. Against a religion based on emotion, the heart, and feeling, Tyndall juxtaposed a religion "founded upon logic," "the religion of the head," and "the religion of the understanding."[48] Whereas the former represented authentic religion, the latter was in reality theology unjustly claiming to be religion. Inner faith was the kernel of religious value within the forms built by theology. Furthermore, Tyndall was not slow to test the worth of orthodox theological doctrine against the superior structure of science. "Science," Tyndall declared, "which is the logic of nature, demands proportion between the house and its foundation. Theology sometimes builds weighty structures on a doubtful base" (*NF,* 13).

The other agnostics also followed Huxley in linking their separation of religion from theology to an affirmation of the basic compatibility between science and religion. Clifford was unsparing in his criticism of Christian theology but treated the true religion of people like Maurice, as well as the religious feeling engendered by cosmic emotion, with tenderness. In a discussion of the conflict between science and theology, Stephen sought a "reconciliation" between science and religion founded on "some deeper principle." He argued that "the sacred images must be once and for all carried fairly beyond the reach of the spreading conflagration, not moved back step by step, suffering fresh shocks at every fresh operation. The radical remedy would be to convey them at once into the unassailable ground of the imagination" (*FP*, 50).

Tyndall likewise believed that science and religion could exist in peaceful harmony. Subjective religious feeling, "as true as any other part of human consciousness," was safe from scientific attack. But any attempt to objectivize emotions, to thrust poetic conceptions into "the region of facts and knowledge," is met by science with hostility. Tyndall pointed out that science therefore makes war only on the scenery, not the substance, of religion. "Let that scenery be taken for what it is worth," Tyndall announced, "as an effort on the part of man to name what by him is unnameable, to express what by him is inexpressible, to bring in short the mystery of life and its surroundings within the range of his capacities, let it be accepted as a symbol instead of asserted as a fact—a temporary rendering in the terms of knowledge of that which transcends all knowledge—and nine-tenths of the 'conflict between science and religion' would cease."[49] Religion, in its subjective dimension and in its articulation through symbol, could be reconciled with the objective facts of science.

Science, according to Tyndall, need not trespass on the sacred mysteries jealously guarded by religion. By talking of nature through the language of science, or in terms of matter, motion, and force, the scientist merely pushed "the mystery back a little" but never plumbed its depths. In fact, Tyndall saw science as being in need of the spiritual sustenance afforded by religion. In his journal he wrote in 1854 that "there are principles in the human heart that cannot be roused by science—principles upon which the culture of science and all other duties depend."[50]

The agnostics, then, saw themselves attacking certain theological doctrines in the interests of religion itself, as well as science.[51] Moore has labored long and hard to demonstrate that, although Huxley used the military metaphor quite frequently, he should not be considered as typical.[52] Moore's estimation of Huxley is faulty, but the view here urged actually reinforces the major thrust of Moore's book, *The Post-Darwinian Controversies.* Those sections in Huxley's work which are often quoted to support the thesis that a state of war existed between science and religion in Victorian England are only examples of Huxley's perception of the antagonism between science and false theology. But Huxley's belief in the genuine harmony of science and religion, rightly conceived, marks him out as typical of the agnostics and the age, and reveals the inadequacy of military historiography.

The tendency to look upon Huxley as the leader of an army of agnostic scientists bent on destroying religion in an inevitable war, and then to proclaim science as the winner, can be undermined if we summarize the paradoxes of this approach in the following manner. The agnostics won the battle in the sense that they de-deified the study of nature, and that is how we do science today. But they just as certainly lost the war in that the agnostics were unable to convince us that they had preserved an authentic religion. We are more likely to pay heed to orthodox Christians such as Wace and Marxist thinkers like Lenin, who claim that the agnostics bore an animosity to religion.

Agnostic Theology and Reactions to Spencer's Unknowable

Although the agnostics tended to champion the cause of true religion and authentic science against the evil influence of Christian theology, they nevertheless believed that a theology of some sort would be acceptable if it met the requirements of science. Huxley was positive that modern science had acted as a beneficial influence in purifying theology and hence religion as well. "If the religion of the present differs from that of the past," Huxley declared, "it is because the theology of the present has become more scientific than that of the past; because it has not only renounced idols of wood and idols of stone, but begins to see the necessity of breaking in pieces the idols built up of books and traditions and fine-spun ecclesiastical cobwebs" (*MR*, 38). A scientific theology was considered to be a legitimate possibility. However, the agnostics reached different conclusions on what survived when theology was submitted to the tests of science. Their disagreement on the content of a valid scientific theology is nowhere more evident than in their varied evaluations of Spencer's Unknowable.

The agnostics were happy to use Spencer's notion of God's unknowableness based on Mansel's antinomies to illustrate the impotence of reason in the realm of religious ideas. They disagreed with Mill's contention that Mansel's assault on intuitive knowledge of God was fallacious, and they followed Spencer in adapting Mansel's approach to their own ends. However, Spencer's theology of the Unknowable did not gain the complete approval of all the agnostics, and some adopted a position that suspended judgment on the existence of God.

In the area of theology there were a variety of shades of opinion. Huxley himself explained why this was so. "The results of the working out of the agnostic principle will vary according to individual knowledge and capacity, and according to the general condition of science," Huxley affirmed. "That which is unproven today may be proven by the help of new discoveries to-morrow. The only negative fixed points will be those negations which flow from the demonstrable limitation of our faculties" (*SCT*, 246, 311). The sole certainties of agnosticism were derived from the demonstrated limitations of the human intellect, which bounded the thought and knowledge of all people. The infinite would eternally transcend the understanding of the finite, and the absolute was forever out of bounds to the relative. But other limits were the result of ignorance, and how far these limits extended depended on the scientific expertise of individuals, their desire to expand their knowledge, and the fund of knowledge available at that point in time. Not surprisingly, then, agnosticism was not monolithic in nature but tended to vary from individual to individual, particularly on the issue of theology.

One of the main reasons for Tyndall's high opinion of "The Unknowable" stems from his acceptance of Spencer's concept of God. In 1866 Tyndall referred to the "Unknowable God."[53] Sprinkled throughout his published works are similar expressions that reveal his theistic leanings. He spoke of the "infinite unknown," the "Inscrutable," "the inscrutable Power . . . in whom we live and move and have our being and our end," and the "Incomprehensible."[54] Even when Tyndall experienced religious doubts he "could by no means get rid of the idea that aspects of nature and the consciousness of man implied the operation of a power altogether beyond my grasp—an energy the thought of which raised the temperature of the mind, though it refused to accept shape, personal or otherwise, from the intellect" (*FS* 2:382).

Whereas Tyndall followed Spencer in vigorously affirming the existence of God, Stephen and Clifford were inclined to suspend their judgment or even tend toward atheism. Stephen did not consider himself to be a disciple of Herbert Spencer, and he rejected Spencer's concept of the Unknowable as a positive consciousness of a God behind appearances. He declared in 1886 that "the unknowable, which lies beyond, is not made into a reality by its capital letter" for to us "it is a mere blank, with which we have nothing to do" (*AA*, 144). Later, in his *English Utilitarians*, Stephen stated: "although I am an 'Agnostic' I cannot accept Mr. Spencer's version of Hamilton's doctrine." Spencer's whole project of presenting a grandiose system of philosophy was viewed by Stephen as premature. Not mentioning Spencer by name, Stephen sarcastically proclaimed that "a man who fancies that he can dictate a complete system to the world only shows that he is arrogant to the verge of insanity." Clifford did not subscribe to the Spencerian deity, nor did he accept any position that gave positive value to the "unreasonable or unknowable."[55]

Although Stephen denied that he was an atheist, his agnosticism sometimes stood slightly to the left of perfectly balanced suspension of judgment on God's existence and nearer to denial. He admitted to Norton in a letter dated 5 March 1876 that his "An Agnostic's Apology" was "of an atheological tendency" (*LLLS*, 287). Stephen believed that the essential value of the Christian faith, as presented by theologians, was retained if humanity, not God, was placed at the center of the religious and moral system (*FP*, 358). Similarly, Clifford, at times, sounded very Nietzschean in his proclamation of the death of God and the coming of the "kingdom of Man" (*LE* 2:285).

Huxley and the Unknowable

Huxley's position on theology is complex, and he has often been misunderstood. Benn and Bury look to Stephen as the purest or most consistent agnostic, because his works contain no traces of the Unknowable.[56] Both scholars tend to place Huxley with Spencer as agnostics who adulterated their thought with metaphysical overtones. However, others, for example Nielsen, turn a deaf ear to Huxley's theistic proclivities and call him and Stephen typical agnostics while excluding Spencer from consideration.[57] Huxley actually displayed affinities with both Stephen and Spencer, for although he attacked the idea of the Unknowable, he nevertheless revealed theistic leanings.

Huxley at first felt comfortable with Spencer's term *the Unknowable*. Writing to Kingsley on 30 April 1863, he referred to "the passionless impersonality of the unknown and unknowable which science shows everywhere underlying the thin veil of phenomena."[58] Like Spencer, Huxley talked of the unknowable behind nature in tones of awe and reverence. Three years later, in his essay "On the Advisableness of Improving Natural Knowledge," he mentioned worshiping "at the altar of the Unknown" (*MR*, 38). During that same year Huxley defended his article on "Science and 'Church Policy,'" recently published in the *Reader*, in a letter to a friend. He stood by his point in the article that theology must stand or fall on its consistency with scientific method, but he asserted "without fear of refutation that there is not a word in that article opposed to any form of belief in a revelation of the Unknowable."[59]

Because of these indications of agreement with Spencer during the sixties, Huxley and Spencer's views of agnosticism were routinely conflated. Even Hutton, who was fairly well informed on the ideas of the agnostics, saw Huxley's position in 1871 as being a mere expansion on Spencer's Unknowable.[60] However, there are strong indications that by the end of the sixties Huxley could no longer subscribe to Spencer's worship of the Unknowable and that one of the reasons for coining the term *agnosticism* was to distance himself from Spencer as well as Positivists, empiricists, and materialists.[61]

For twenty years after coining the term *agnosticism* Huxley refused to mention the Unknowable in his published works. Then in 1889 he announced that he did not "care to speak of anything as 'unknowable,'" appending a note of confession to this passage four years later that "long ago, I once or twice made this mistake; even to the waste of a capital 'U'" (*SCT,* 311). Huxley's main concern was to avoid a reference to God as the unknowable (or even worse the Unknowable), for this implicitly granted God ontological status through some sort of act of consciousness, and hence amounted to knowing the unknowable.

In his private correspondence Huxley was less tender in his criticism of Spencer. In a revealing letter to Gould he accused Spencer of succumbing to the crudest idolatry in his worship of "negative abstractions," and he pointed to the vast gulf between agnosticism and Spencer's position:

> As between Mr. Spencer and myself, the question is not one of a "dividing line," but of an entire and complete divergence as soon as we leave the foundations laid by Hume, Kant, and Hamilton, who are my philosophical forefathers. To my mind, the "Absolute" philosophies were finally knocked on the head by Hamilton; and the "Unknowable," in Mr. Spencer's sense, is merely the Absolute *redivivus,* a sort of ghost of an extinct philosophy, the name of a negation hocus-pocussed into a sham thing. If I am to talk about that of which I have no knowledge at all, I prefer the good old word *God,* about which there is no scientific pretense.[62]

Huxley's frustration at seeing his own brainchild agnosticism perverted by Spencer can be read between the lines of this private letter.

His indignation finally spilled over into his published work in 1895, when he intended to air his basic differences with Spencer in public. Ironically, the second half of "Mr. Balfour's Attack on Agnosticism" was never published, since Huxley died before completing it, and therein were the strong statements on his disagreement with Spencer. Huxley recounted his mixed feelings about Spencer's *First Princi-*

ples when it first appeared. Although he welcomed Spencer's critical positions warmly, "even then Mr. Spencer appeared to me to be disposed to travel along the path—by which, as I conceive, Hamilton had been led astray—further than I was. And in the forty-three years which have elapsed the divergence of opinion thus marked has unfortunately become greater and greater, until now we are speculatively (I hope in no other way) poles asunder."[63] Huxley had moved closer to Mill during all those years. In the first part of "Mr. Balfour's Attack on Agnosticism," which was published in 1895 in the *Nineteenth Century*, Huxley repeated Mill's criticism of Hamilton that under the guise of faith the Scottish philosopher had inconsistently admitted all that he had ruled out as unknowable (535). The point was, of course, equally applicable to Spencer.

It is not surprising that Huxley's version of agnosticism was often confused with Spencer's doctrine of the Unknowable. Their close personal association was well known. Equally evident was their agreement on the importance of science and the inadequacy of Christian orthodoxy, and their shared opinions on the limits of human knowledge. Both of them called themselves agnostics. Although Huxley became more and more alienated from Spencer's position as the years passed, he was discreet. He did not correct men such as Hutton when they conflated the two types of agnosticism. Huxley recognized that he was therefore partly to blame for the confusion. "I have long been aware of the manner in which my views have been confounded with those of Mr. Spencer," Huxley told Gould in 1889, "though no one was more fully aware of our divergence than the latter. Perhaps I have done wrongly in letting the thing slide so long, but I was anxious to avoid a breach with an old friend."[64] But in 1889 a breach between Huxley and Spencer had occurred, ostensibly over their controversy in the *Times* on land nationalization. Huxley had actually raised questions about the validity of Spencer's whole a priori approach; this obviously implied disagreement in nonpolitical areas, including religious thought. The quarrel began in November of 1889 and continued for four years, much to the consternation of close friends like Tyndall. Huxley's letters to Gould were written after the squabble had begun, and undoubtedly his hostility toward Spencer loosened his tongue on their differing views of agnosticism.[65]

Although Huxley attacked Spencer's idolatrous worship of the Unknowable, he still retained elements of a theistic position and allowed for the validity of some type of theology. In "A Liberal Education, and Where to Find It" (1868) Huxley declared that our life and happiness depend upon our knowledge of the rules of a game infinitely more difficult than chess. "The chess-board is the world," Huxley affirmed, "the pieces are the phenomena of the universe, the rules of the game are what we call the laws of Nature. The player on the other side is hidden from us. We know that his play is always fair, just and patient." Though "hidden" the opposing player could be seen as "a calm, strong angel who is playing for love . . . and would rather lose than win" (*SE*, 82). Whereas in 1863, Huxley had talked of the passionless impersonality of the unknown and unknowable, here he pictures a passionate angel as underlying phenomena. If ultimate reality is beyond human knowledge as Huxley claimed, then it is hard to see how he escapes the charge of idolatry or anthropomorphism which he leveled at Spencer. Huxley's inconsistency in this matter is a prime illustration of his rather ambivalent attitude toward theism.

Although Huxley maintained that theology is not the whole of religion, and although he actually saw the reverence for the ethical ideal as a more important element in religion, he nevertheless affirmed that a proper theology was a valid part of religion. A scientific theology was a real possibility:

> But it is at any rate conceivable, that the nature of the Deity, and his relations to the universe, and more especially to mankind, are capable of being ascertained, either inductively or deductively, or by both processes. And, if they have been ascertained, then a body of science has been formed which is very properly called theology. Further, there can be no doubt that affection for the Being thus defined and described by theologic science would be properly termed religion; but it would not be the whole of religion. (*SE*, 394–95)

Huxley was sincere when he exclaimed that "if the belief in a God is essential to morality, physical science offers no obstacle thereto" (*EE*, 143).

Huxley's theistic yearnings were rarely displayed to the public who read his essays and heard him speak. But to friends he was more willing to open up and speak from the depths of his soul. In 1879 Fiske dropped by the Huxleys to say good-bye before he left England to return to America. Huxley took Fiske up to his study where they sipped a glass of toddy and puffed on cigars. "Then Huxley and I got into a solemn talk about God and the soul," Fiske remembered, "and he unburdened himself to me of some of his innermost thoughts—poor creatures both of us, striving to compass thoughts too great for the human mind." During one of his many intimate conversations with Wilfrid Ward, Huxley remarked that the Christian definition of theism, "faulty and incorrect" as it may be, "is nearer the truth than the creed of some agnostics who conceive of no unifying principle in the world."[66]

Spencer and His Disciples

Huxley realized that his complex position on theology could be open to misinterpretation, particularly in light of his earlier endorsement of Spencer's Unknowable. He became more concerned in the eighties when he perceived that not only was his agnosticism conflated with Spencer's, but worse still, Spencer was being considered the leading agnostic, and his dangerous idolatry was infecting the thought of the agnostic rank and file. Huxley wrote to Charles Albert Watts in 1883 that "until now 'agnostics' are assuming the character of a recognised sect," and he indicated his dissent from the movement by declaring that if there were called a "General Council of the Church Agnostic very likely I should be condemned as a heretic." An agnostic heretic, Huxley recognized, made a poor leader, and he publicly gave up the honorary title of pope which Hutton had earlier bestowed. He admitted in 1889 that if a sect of agnostics existed, "I am not its acknowledged prophet or pope." In 1892 Huxley claimed that he did not desire to imitate Comtists by assuming the position of "master of a school, or leader of a sect, or chief of a party." He added that history had taught him to see that schools or parties usually perpetuate all that is worst and feeblest in the founder's work. Huxley obviously feared that some of the agnostics had already done just that to his original coinage.[67]

The pope of agnosticism actually was Spencer. Huxley was not widely considered to be the leader or even founder of agnosticism during the Victorian era. Upon reading an article entitled "Modern Skepticism" in the *Scribner's Monthly* for 1873 Huxley wrote Tyndall and joked about how he had emerged virtually untouched by the author's criticisms while Spencer and Tyndall were attacked as the head sceptics. "I come in only *par parenthèse*," Huxley pointed out to Tyndall, "and I am glad to see that people are beginning to understand my real position, and to separate me from such raging infidels as you and Spencer."[68] Other contemporaries who needed a typical agnostic to attack chose Spencer over Huxley. In an article called "The Coryphaeus of Agnosticism" (1882), the *Month* treated Spencer as the chief agnostic, while Harrison referred to him as "the most important leader of the pure Agnostic school" in 1889.[69]

Spencer's "Unknowable" was a lucid, systematic statement of the agnostic position presented in the early sixties, and was considered by many to be an official handbook. Huxley did not have the advantage of a comprehensive exposition of his view of agnosticism until 1889. His *Hume* included a brief discussion of agnosticism but it was presented as Hume and Kant's doctrine, not as his own, and he made no claims to authorship of the term. Spencer's "Unknowable" therefore stood as the main textbook of agnostic theory until 1889, when Huxley finally elaborated on the original meaning of the term he had coined over twenty years before. But Huxley's major articles focusing on agnosticism came rather late in the development of the new creed.[70]

Some of the agnostics were not impressed with Huxley's articles "Agnosticism," "Agnosticism: A Rejoinder," and "Agnosticism and Christianity," all published in 1889 (in February, March, and April respectively in *Nineteenth Century*). Tyndall wrote to Hirst on 15 April 1889: "I have read Huxley's articles. That on miracles especially I thought of little use. It was hacking a dead horse. In these matters I am in favour of the decency of slow decay." Hirst's response on 22 April, indicated his agreement with Tyndall. Hirst added, "you and I were Agnostics long before our friend Huxley invented the word." Hirst's journal reveals that he was reading all three of Huxley's pieces and that he was not enthused by the whole series. On 12 June he records that he read the last article and "to tell the truth I became weary of it."[71]

Huxley is sometimes classed as leader of the agnostic movement because he coined the term *agnostic*, but a number of agnostics were not actually aware of Huxley's prowess as neologist. Huxley announced his claim in the first article in the series, "Agnosticism." Hirst was surprised. "He mentions a fact previously unknown to me," Hirst wrote in his journal, "viz that Agnosticism was the name given by (Huxley) himself to his own creed."[72] It is unclear when Stephen first discovered that Huxley had coined the word *agnostic*, but he was unaware of this fact when he published "An Agnostic's Apology" in 1876.[73]

Although the major agnostics, Tyndall, Clifford, and Stephen, did not see Huxley as their leader, they also worked independently of Spencer. Even Tyndall, who was closest to Spencer in his conception of agnosticism, was not inclined to look to Spencer for approval. However, this was not the case with less well-known agnostics such as Gould, Bithell, and Laing, who attempted to popularize science and agnosticism by putting forward the agnostic position in a systematic fashion during the eighties and nineties. They tended to view Spencer as their master and the Unknowable as their deity.

Frederick James Gould (1855–1938) worked as a village schoolmaster and London Board school teacher from 1877 to 1896. He became involved in the Ethical Movement in the late nineties, and in 1899 was made secretary of the Leicester Secular Society, a post he held for nine years. Thereafter, Gould became immersed in lecturing on moral education, and with government backing he toured India in 1913 and the United States in 1914. From 1919 to 1927 he was appointed Honorary Secretary of the International Moral Education Congress.[74]

Gould passed through an agnostic phase in the late eighties and early nineties before going on to work in the Ethical Movement and organized secularism.

On 23 December 1889 Gould wrote to Huxley for information to help him prepare a pamphlet on agnosticism. After reading Huxley's *Hume* and the first article on agnosticism published in February in the *Nineteenth Century*, Gould perceived a gap between Spencer and Huxley which he had not previously recognized, "namely, that you do not go so far as a *positive* affirmation of the Unknowable Noumenon." In response to Gould's question Huxley wrote the interesting criticism of Spencer already discussed. Gould wrote back on 2 January 1890 and sympathized with Huxley's embarrassment in regard to the confusion surrounding agnosticism. But Gould pointed out that the muddle was understandable since both Huxley and Spencer started with the same critical attitude. "I suppose the best way out of the dilemma," Gould suggested, "would be to invent another name for Spencerian Agnosticism" (ICST-HP 17:107, 109).

But when he published his pamphlet *Stepping-Stones to Agnosticism* (1890), Gould's depiction of agnosticism was unashamedly Spencerian. Belief in a God was laid down as an agnostic principle on the basis that Spencer's "doctrine of the Unknowable is assented to by so many professed Agnostics." In a section on the subject of the limits of human knowledge he presented a short summary of Spencer's "The Unknowable" and referred readers to that same work for a more detailed study.[75]

Huxley's statement that agnosticism varied with each individual was never illustrated more strikingly than in the cases of Laing and Bithell. Like Gould, they considered themselves to be disciples of Spencer, and they stressed the theistic and religious dimension of agnosticism. More than Gould, they presented certain idiosyncratic doctrines of their own as necessary components of any theory of agnosticism.

Richard Bithell (b. 1821), B.Sc. (London University) and Ph.D. (Gottingen), Fellow of the Institute of Bankers and a member of the London Dialectical Society, was the first to see the need for a comprehensive statement of agnosticism in the eighties which would update Spencer's "The Unknowable."[76] In his *Creed of a Modern Agnostic* (1883) he asserted that he was "not aware of any one [who] has, as yet, stated his views on Agnosticism as a definite creed."[77] He willingly expressed his sense of obligation to Spencer in his task of presenting a system of agnostic theory, writing: "I am indebted to Mr. Spencer's works more than to those of any other writer." Referring to "that large body of Theistic Agnostics" who recognized "the existence and activity of a Supreme Power," Bithell insisted that the agnostic deity be Spencer's "Unknowable." "We not only can worship the Unknowable," Bithell declared, ". . . it is the only proper object of supreme worship."[78] Spencer's reverence for the Unknowable, by and large kept under control and enunciated in detached, passionless terms, was transformed by men such as Bithell who did not hesitate to indulge their feelings of adoration for their deity.

A barrister, an official of the Board of Trade, three times a member of Parliament for the Liberals (1852, 1868, 1873), and a successful chairman of the London, Brighton, and South Coast Railway, Samuel Laing (1812–1897) became an author at age seventy and produced a series of books which garnered him an influence with the general public almost equal to that of the chief thinkers of the day.[79] Robertson remarked that although Laing was not a man of genius "he perhaps converted more men to rationalism in the 'eighties and 'nineties than any other British publicist did by book-work in the 'seventies."[80] Laing's *Modern Science and Modern Thought* (1885) was a bestseller, second only to Haeckel's *Riddle of the Universe* in terms of total sales of cheap reprint editions published by the Rationalist Press Association in 1903.[81]

In the preface to his *Modern Zoroastrian* (1887), Laing reported that for the scientific portion of the book he was indebted to Darwin, Huxley, and Haeckel. "For the religious and philosophical speculations I am myself responsible; for, although I have derived the greatest possible pleasure and profit from Herbert Spencer's writings, I had arrived at my principal conclusions independently before I had read any of his works" (x). But although Laing wished to be seen as an example of a self-help agnostic, his position clearly resembled Spencer's, rather than Huxley's, brand of agnosticism. "Directly we pass beyond the boundary of such knowledge as really can be known by human faculty," Laing affirmed, "and stand face to face with the mystery of the Great Unknown, we can only bow our heads with reverence and say with the poet, Behold, I know not anything."[82] Laing's approval of what he called "Christian Agnosticism," and his attempt to outline the "Religion of the Future," clearly mark him out as one who underscored the religious quality of agnosticism.[83]

Both Laing and Bithell strike the modern reader as faddists in their rigorous affirmation of doctrines that seem strangely inconsistent with their agnosticism. Bithell's pet theory concerned the existence of a "spiritual body," which he claimed provided scientific evidence for the notion of immortality in comparison to the unscientific Christian idea of the soul.[84] Laing's favorite idea involved the promulgation of Zoroastrianism, the ancient Persian religion that recognized the existence of a force for evil which would ultimately be defeated by a force for good. "Now of all the religious hypotheses which remain workable in the present state of human knowledge," Laing maintained, "that seems to me the best which frankly recognises the existence of this dual law, or law of polarity, as the fundamental condition of the universe, and, personifying the good principle under the name of Ormuzd, and the evil one under that of Ahriman, looks with earnest but silent and unspoken reverence on the great unknown beyond, which may, in some way incomprehensible to mortals, reconcile the two opposites, and give the final victory to the good." The advantage of Zoroastrianism, to Laing, lay in its refusal to conceive of God as omnipotent, since this absolved God of responsibility for evil and jived well with the suffering and waste implied by evolutionary theory.[85]

To Huxley, Laing's eccentricities epitomized a disturbing new element in the development of agnosticism. A vast number of agnostics, with Spencer as their prophet, had built a new church or temple which mimicked a number of unhealthy Christian practices. When Laing compiled a list of eight articles of the agnostic creed, the second of which affirmed the existence of an "inscrutable First Cause" and the eighth of which proclaimed polarity as "the great underlying law of all knowable phenomena," Huxley could not restrain himself.[86] His article "Agnosticism" is in part an attack on Laing and a means of defending himself from Wace and other orthodox Christians. Huxley wrote that agnosticism "has been furnished with a set of 'articles' fewer, but not less rigid, and certainly not less consistent than the thirty-nine," and he dismissed Laing's articles on the grounds that agnosticism is a method and not a creed (*SCT*, 209, 245). Turning then to Laing's theory of polarity in particular, Huxley deemed it less clear than the Athanasian creed, and he compared it to the *Naturphilosophie* from which he had suffered in his youth. "For many years past, whenever I have met with 'polarity' anywhere but in a discussion of some purely physical topic, such as magnetism, I have shut the book," Huxley told his readers. "Mr. Laing must excuse me if the force of habit was too much for me when I read his eighth article" (*SCT*, 247). Perhaps Huxley's harsh criticism was misplaced, for it was because of his earlier tolerance of varied forms of agnosticism that Spencer and his disciples came to define it so differently.

Whatever their view on theology, be it Christian or agnostic, Huxley, Stephen, Tyndall, and Clifford agreed that the theologian attempted to embody religious emotion in the language of the intellect. The source of religious feeling was largely the sense of awe produced by contact with nature. Keenly aware of the wondrous order in the natural world, the original agnostics often adopted the manner of the enthusiastic natural theologian. But was their new religion of science in conflict with their agnostic principles? Did the limits of knowledge act as an obstacle to the justification of scientific principles that grounded their version of natural theology? To answer these questions we will examine the agnostics' worship of nature in the next chapter.

Chapter Six

THE NEW NATURAL THEOLOGY AND THE HOLY TRINITY OF AGNOSTICISM

> *Naturalism, we find, though rejecting materialism, abandons neither the materialistic standpoint nor the materialistic endeavour to colligate the facts of life, mind, and history with a mechanical scheme. But the compact of Naturalism with Agnosticism . . . costs Naturalism, as it turns out, its entire philosophical existence. In order to be free of 'metaphysical quagmires' such as the ideas of substance and cause, it is led to reject the reality not only of mind, but even of matter; and in this state of ideophobia must collapse, for lack of the very ideas it dreads.*
>
> JAMES WARD

Agnosticism was originally conceived of by Huxley as a powerful weapon to be used by science against the false pretensions of orthodox theology. Ironically, the marriage between agnosticism and scientific naturalism did not work out. As formulated by Huxley, Tyndall, Clifford, Stephen, and Spencer, the agnostic position was peculiarly vulnerable in areas that could only embarrass such staunch defenders of the value of natural science. Applying the Manselian idea of the limits of knowledge to the natural world proved to be as destructive to scientific naturalism as to orthodox Christianity. The sceptical element of the Victorian agnostics' thought made it difficult for them to demonstrate the reality and validity of the crucial scientific principles of the universality of the law of causation, the uniformity of nature, and the existence of an objective, external, natural world. In a sense, these three scientific axioms became articles of faith, for the agnostics could no more justify certainty in their existence than orthodox Christians could scientifically prove the actuality of the Son, the Father, and the Holy Ghost.

The agnostic holy trinity not only justified faith in the methods of natural science, it also provided the major axioms upon which the agnostics erected a new version of natural theology. The agnostic dedication to the worship of nature is most noticeable in their fascination with Carlyle's natural supernaturalism, their love of the Alps, and their overwhelming sense of the order in nature. But since their holy trinity could be justified by faith alone, the status of agnostic scientific theology differed little from that of Christian theology.

Natural Supernaturalism and the Worship of Nature

"The agnostic doctrines," Carlyle once told J. A. Froude, "were to appearance like the finest flour, from which you might expect the most excellent bread; but when you came to feed on it you found it was powdered glass and you had been eating the deadliest poison."[1] But despite Carlyle's low opinion of agnosticism, Tyndall, Huxley, and even Stephen and Clifford were influenced in varying degrees by his natural supernaturalism. In 1866 Tyndall and Huxley were among four to receive doctorates from the University of Edinburgh. Since Thomas Carlyle was to be installed as rector of the university at the same ceremony, Tyndall took charge of Carlyle on the trip to Edinburgh. It was a difficult journey for Carlyle, who was seventy-one. Tyndall personally supervised Carlyle's eating and sleeping schedule and succeeded in staving off Carlyle's old enemy insomnia, which could have destroyed his speech as the new rector. Carlyle later wrote to his wife that "Tyndall's conduct to me has been loyalty's own self; no loving son could have more faithfully watched a decrepit father: in fact I shall not forget it" (*LWJT,* 121–22).

Encapsulated in the story of Tyndall and Huxley's trip to Edinburgh with Carlyle is the debt scientific naturalism owed to natural supernaturalism. The agnostics were attracted to Carlyle's stress on work, his call for an aristocracy of talent to enact social reform, his attack on the Anglican clergy who prevented the reconstruction of society, his optimistic vision of the world, and his view of religion as wonder, humility, and work within a divine universe.[2] It is the agnostic sense of the divinity of nature which is most relevant for our study of the religious dimension of agnosticism. In the agnostic worship of nature we can find traces of Romanticism and *Naturphilosophie,* in particular, the thought of Goethe, Schiller, and Fichte.

The agnostics conceived of nature as a living, organic entity rather than as a mechanical, inorganic thing composed of dead atoms. In 1856 Huxley declared that "in travelling from one end to the other of the scale of life, we are taught one lesson, that living nature is not a mechanism but a poem; not a mere rough engine-house for the due keeping of pleasure and pain machines, but a palace whose foundations, indeed, are laid on the strictest and safest mechanical principles, but whose superstructure is a manifestation of the highest and noblest art."[3] In 1869 Huxley translated a poem by Goethe entitled "Nature" for the first issue of the journal *Nature*, and he remarked that "long after the theories of the philosophers whose achievements are recorded in these pages, are obsolete, the vision of the poet will remain as a truthful and efficient symbol of the wonder and the mystery of Nature."[4] Above the transitory theories of science Huxley placed the eternal character of art. Nature was a poem to Huxley, more truthfully dealt with by poetry than by science, due to its beauteous, mysterious, and angelic quality. Huxley once told Wilfrid Ward: "one thing which weighs with me against pessimism and tells for a benevolent Author of the Universe is my enjoyment of scenery and music. I do not see how they can have helped in the struggle for existence. They are gratuitous gifts."[5]

Clifford talked of "cosmic emotion," or a sense of "awe, veneration, resignation, submission" felt "in regard to the universe or sum of things, viewed as a cosmos or order." According to Clifford there existed two kinds of cosmic emotion, one corresponding to each of the worlds we experience, the macrocosm or the universe that surrounds and contains us, and the microcosm, which is the world of our own souls. Clifford believed that Kant had expressed a "special form of each of these kinds of cosmic emotion" in his sentence "Two things I contemplate with ceaseless awe; the stars of Heaven, and Man's sense of Law" (*LE* 2:253–54). Stephen was sometimes moved by nature to express similar feelings. "Carlyle has been," Stephen stated, "to some of us, the most stimulating of writers, just because he succeeded in expressing, with unsurpassed power, the emotion which I must be content with indicating—the emotion which is roused by sudden revelations of the infinitudes, the silences and eternities that surround us." The emotion of wonder Stephen saw, like Huxley, as expressed more appropriately in poetry. Stephen declared that "to us a star is a signal of a new world; it suggests universe beyond universe; sinking into the infinite abysses of space; we see worlds forming or decaying and raising at every moment problems of a strange fascination. The prosaic truth is really more poetical than the old figment of the childish imagination."[6]

But it was Tyndall who often outdid all of the other agnostics in his expressions of wonder on the divine element in nature. To Tyndall even matter was mystical and transcendental.[7] "If the power to build a tree be conceded to pure matter," Tyndall declared, "what an amazing expansion of our notions of the 'potency of matter' is implied in the concession!" Upon reading the final page of Huxley's "On the Physical Basis of Life," where his friend discussed how an ontological materialism may paralyze "the energies and destroy the beauty of a life," Tyndall wrote Huxley and applauded his point. To Tyndall, Huxley was affirming his notion of a redefined materialism. "I hope some day to see you develope the last page," Tyndall told Huxley, "and particularly the last paragraph of the 'Physical Basis of Life.' I probably am the only living man who sees the meaning shadowed forth in that page and paragraph!!"[8]

Nature contained a mysterious spiritual element for Tyndall which demanded reverence and worship. After viewing a magnificent sunset in June 1850, Tyndall concluded that surely some "principle" permeates nature:

> Who created it? What is it? The soul yearns over the mystery, retires baffled, but will try again. Encompassed by such thoughts, revelation seems common-place, for who-ever listens with reverential ear will not he also detect the spirit voice speaking in melody to his soul; supernal whispers which, fitly uttered, would be as good and true as any revelation of them all. My experience is precisely that of the most orthodox christian.[9]

Such devout feelings toward nature mark out Tyndall as a quasi pantheist.[10] Like the other agnostics, Tyndall would be best described as a scientific natural supernaturalist.

Stephen, Tyndall, and the Alps

For some agnostics the almost divine quality of nature touched them deepest when they were on the slopes of the Alps. In a review of Tyndall's *Glaciers of the Alps* (1860), Huxley noticed that some supporters of science displayed a version of muscular Christianity in their romance with mountain climbing. "An ingenious speculator," Huxley affirmed, "indeed, might develope the parallel between the ecclesiastical and the scientific sects to a great length. The difficulties and obstacles which the Alps present to a scientific explorer are of a very similar order to those which a poaching village on the borders of the New Forest, or a parish in the Potteries, offer to a reforming rector."[11] Huxley's observation could be applied to Leslie Stephen and John Tyndall, both of whom were avid lovers of the Alps with important first ascents under their belts.

Stephen openly admitted that he was a fanatic when it came to mountain climbing. "I believe that the ascent of mountains forms an essential chapter in the complete duty of man," Stephen announced, "and that it is wrong to leave any district without setting foot on its highest peak."[12] Stephen was infected by mountain fever in 1857, the same year the Alpine Club was first founded.[13] In 1858 he joined the

Club (as did Tyndall) and began to play a leading role in its activities. He was elected president in 1865, and he edited the *Alpine Journal* from 1868 to 1871. His *Playground of Europe* (1871) was immensely popular, and it reinforced the enthusiasm of the mid-Victorians for the Alps.

In addition to appreciating the challenge that mountain climbing offered as a test of character, Stephen often spoke of the Alps in tones of religious awe. Harrison observed that "the Alps were to Stephen the elixir of life, a revelation, a religion."[14] Indeed, Stephen talked of how "the Oberland is to me a sacred place," and he singled out one mountain, the Wengern Alp, as "a sacred place—the holy of holies in the mountain sanctuary."[15] From the summit of the Rothorn Stephen could see the magnificent Weisshorn, and he was "absorbed in the worship of this noblest of Alpine peaks," but the whole of the Oberland beckoned him in 1877 "to worship there again."[16]

Stephen worshiped the Alps because they evoked a special feeling and emotion akin to a religious experience. Joking about inventing a new idolatry, Stephen vowed to prostrate himself before the Alps, whose gigantic masses suggested to him "some shadowy personality" who spoke in mystical tones "at once more tender and more awe-inspiring than that of any mortal teacher." The incomparably exquisite beauty of the winter Alps "belonged to the dream region in which we appear to be inspired with supernatural influences," they elicited "pure undefined emotion" that seemed "to belong to the sphere of the transcendental." Sometimes the Alps impressed upon Stephen our oneness with nature and the universe, at other moments they shocked him out of his complacency when their eternal and infinite quality reminded him of our petty and ephemeral existence. The Alps were at once melancholy and exhilarating, for they instilled an awestruck humility in the face of "the indomitable force of nature to which we are forced to adapt ourselves" while simultaneously inspiring us with visions of the beauty of nature.[17]

Perhaps Stephen had Tyndall in mind when he charged that modern writers could not do justice to the beauty of the Alps when they insisted on including accounts of scientific experiments and discoveries in their descriptions of Alpine scenery.[18] In fact, it was criticism of those who mixed science with mountain climbing which led Tyndall to resign from the Alpine Club in 1862 even though he was vice-president and was in line for the presidency.[19] Stephen's remarks in a speech during an Alpine Club dinner in 1861 apparently hastened Tyndall's withdrawal (*LWJT,* 390).

Stephen had misread Tyndall, as the latter did not allow his interest in glaciers and in other scientific experiments performed while on a climbing expedition to interfere with his worship of the Alps. To Tyndall the Alps were a tonic to cure the ill effects of London life. At a dinner party in 1869 he was overheard by the American Charles Eliot Norton (1827–1908) to exclaim in his Irish brogue, "Ah! the mountain tops, 'I is there that man fales himself nearest the devine. I always sakes the mountain tops for relafe from the tile and care of the wurrld."[20]

For Tyndall, climbing a mountain was like struggling with a spiritual or religious difficulty. Huxley referred to Tyndall as the "mightiest evangel" of that "sect of muscular philosophers whose best-known church is the Alpine Club."[21] The obstacles presented by crevasses, glaciers, and dangerously steep mountain faces were to be met in a strong, vigorous fashion in order to win the battle with the sin of laziness and gain new spiritual insight. Once the peak was reached and the battle won, one gained an invaluable sense of self-respect as well as the pleasure of perceiving the harmony and beauty of God's creation. Tyndall's descriptions of the Alps sometimes read like a form of religious experience.

Tyndall could not resist launching into a rhapsody on the splendor of nature when he was high upon Alpine peaks. During his expedition to the Mer de Glace in 1859 Tyndall encountered a mountain summit which was bathed in a red light. "The adjacent sky wore a strange and supernatural air; indeed there was something in the whole scene which baffled analysis, and the words of Tennyson rose to my lips as I gazed upon it:—'God made Himself an awful rose of dawn.' "[22] The Jungfrau he described as "consecrated ground," and the summit of the Weisshorn drew from his lips an ode to "the transcendent glory of Nature" in which "I entirely forgot myself as man."[23] A sense of the order and oneness of nature, of which we, too, are a part, often overwhelmed Tyndall during his mountain climbing expeditions.

Two mountains held special significance for Tyndall, the Matterhorn and the Weisshorn. Tyndall's contest with the Matterhorn, considered the most difficult ascent of the Alps and not conquered until 1865, finally ended in 1868 when he succeeded in reaching the summit after several failed attempts. During the unsuccessful expedition of 1862 Tyndall spoke of the Matterhorn as "our temple, and we aproached [*sic*] it with feelings not unworthy of so great a shrine." When Tyndall triumphed over the Matterhorn in 1868 the worn aspect of its crags saddened him, as they suggested "inexorable decay." But this started him thinking of days past when the mountain was at full strength, of its origin, and even the birth of the universe in its entirety. The Matterhorn evoked in Tyndall very strong emotions and thoughts, and it raised a series of cosmic questions. Did the nebulous haze which was the source of all material things "contain potentially the *sadness* with which I regarded the Matterhorn? Did the *thought* which now ran back to it simply return to its primeval home? If so, had we not better recast our definitions of matter and force? for if life and thought be the very flower of both, any definition, which omits life and thought must be inadequate, if not untrue." The Matterhorn represented for Tyndall his whole transcendental notion of matter because its material aspect impressed upon those who viewed it the necessity of grasping the spiritual meaning of nature.[24]

The majesty of the Weisshorn left Tyndall speechless. In his published account of the ascent he tells his readers that he opened his notebook to make a few observations but soon relinquished the attempt. "There was something incongruous," Tyndall believed, "if not profane, in allowing the scientific faculty to interfere where silent worship was the 'reasonable service.'" However, in his journal Tyndall was less reserved. From the vantage point of the top of the Weisshorn Tyndall could see a number of mountains, the peaks of which were illuminated in such a way that "they seemed hung in heaven like a chain of opals; fit to form a necklace for their Almighty Maker."[25]

Tyndall felt that his scientific work and his love of the Alps were natural complements. In terms of importance he placed his Alpine writings on a par with his strictly scientific volumes. In the preface to his *Hours of Exercise in the Alps* (1871) Tyndall remarked that "a short time ago I published a book of 'Fragments,' which might have been called 'Hours of Exercise in the Attic and the Laboratory'; while this one bears the title of 'Hours of Exercise in the Alps.' The two volumes supplement each other, and taken together, illustrate the mode in which a lover of natural knowledge and of natural scenery chooses to spend his life" (v). Tyndall was not very different at all from his fellow agnostic Stephen in his attitude toward the awe-inspiring Alps.

The New Natural Theologians

In 1889 Huxley told the readers of the *Nineteenth Century* that his aim was to rouse his countrymen out of their dogmatic slumbers, but not in order to see who would win in a contest between scientist and theologian. "The serious question," Huxley maintained, "is whether theological men of science, or theological special pleaders, are to have the confidence of the general public" (*SCT*, 270). Science, to Huxley, was not inevitably opposed to all theology. Although his view of theology as a science permitted him to attack the false theology of ecclesiasticism without disturbing religion, it also served to allow Huxley to create a new theology drawn from the discoveries of science. The majority of the agnostics followed suit, and they must be classed as one of the important schools of thought which perpetuated the tradition of natural theology developed in the first half of the nineteenth century, albeit in a modified form. The god that nature seemed to intimate, and which the agnostics sensed through their religious feelings, could be studied by the intellect through science.

The heart of the new natural theology was an emphasis on the ability of science to uncover the order in nature through an empirical study of the physical world. Huxley not only believed that science was capable of revealing the natural order, he also affirmed in 1887 that the prime object of the physical inquirer "is the discovery of the rational order which pervades the universe."[26] The agnostics looked upon nature as a whole and found it a world of order.

As scientific naturalists, the agnostics believed that natural phenomena could be explained without recourse to a transcendental cause who intervened directly from time to time. The natural order was susceptible of empirical study, but the majority of the agnostics were convinced that this order revealed the existence of a mysterious deity whose will was manifested in the laws of nature. Science, then, was still a study of the ways of God, for it showed that God worked so artfully through secondary causes that his immediate intervention was unnecessary. To underline his point that there was a line of continuity between early-nineteenth-century natural theology and the later scientific naturalism, Young perceptively points out that "the view of God changed from a natural theology of harmony in nature and society (with direct appeals for explanation of their order), to a Deity identified with the self-acting laws of nature."[27] The deity of agnosticism was virtually synonymous with the laws of nature or the natural order. Recent scholarship has suggested that this was also Darwin's view of God's relation to nature, at least while he wrote and published the *Origin*.[28]

Clifford's "cosmic emotion" represented his idea of the appropriate response to the divine order in nature. He discussed in an approving tone the view of nature in *The Golden Verses* of the fifth-century Neoplatonic philosopher Hierokles. Here indeed was a conception of nature which was fit for evoking a feeling of cosmic emotion, for *The Golden Verses* taught us "to look upon Nature as a divine Order or Cosmos, acting uniformly in all of its diverse parts; which order, by means of its uniformity, is continually educating us and teaching us to act rightly" (*LE* 2:267).

Stephen also tended to deify the natural order. He maintained that "the scientific or Darwinian view cannot deny positively the existence of a plan, or prove the world to be 'accidental,' or irrational," but it demands that "the plan, if there be a plan, can be known only by observation."[29] But when he looked at the world Stephen was impressed with its orderliness. "The theologian," Stephen wrote, "agrees with the man of science in admitting that we are governed by unalterable laws, or, as the man of science prefers to say, that the world shows nothing but a series of invariable sequences, and coexistences." Stephen balked at the next step that the theologian made, the movement from natural law to a transcendental lawgiver. The true scientist saw nothing behind natural law "but impenetrable mystery" that could be regarded as divine, yet immanent (*FP*, 351–53). God, viewed as an immanent lawgiver, was "not an external ruler" but "an all-pervading essence." Stephen was critical of the ordinary Briton who was "so inquisitive that he insists upon knowing whether the word God is to be applied to a being who will interfere, more or less, with his life, or is merely a philosophical circumlocution for the unvarying order of nature" (*FP*, 59). Stephen opted for the latter alternative. He argued for the elimination of the supernatural and an identification of "the Divine element" with "the natural order" (*AA*, 74).

Like Stephen, Tyndall's naturalistic interpretation of nature did not prevent him from finding an order in the physical world. Tyndall objected to Paley's form of natural theology because it presented a false and irreligious picture of God's relation to nature.[30] In 1849 Tyndall noted in his journal that he had read Paley's *Natural Theology* and was "willing enough to be convinced upon the subject of which Paley writes, but what he says is insufficient for this end. The Great Spirit is not to be come at in this way; if so, his cognition would only be accessible to the scientific and to very little purpose even here." Paley's God was "an omnipotent mechanic detached from his work," and Tyndall told Hirst that "the theist of Paley's class is I believe intrinsically the same as the atheist." In opposition to Paley's God, which was conceived of as outside the world, Tyndall believed "with Carlyle [that] the universe is the blood and bones of Jehovah—he climbs in the sap of trees and falls in cateracts."[31]

Tyndall's immanent deity was behind the natural order that so delighted him. The scientist's job was to uncover the order that underlay the apparent chaotic state of nature. Scattered throughout Tyndall's writings are expressions of wonder at the beauty and harmony of nature. Speaking of molecular architecture Tyndall remarked that "its beauty would delight and astonish you." In studying the crystalline form of ice, snow, and frost, located in "a region withdrawn from the inattentive eye, we find ourselves surprised and fascinated by the methods of Nature." It would be difficult to find many orthodox natural theologians who outdo Tyndall in his enthusiastic descriptions of the beauty and wonder of the physical world. He was strongly attracted both to the microworld of the atom as well as the macroworld of the Alps. Tyndall was no doubt sincere when he recommended to his fellow countrymen that they spend the sabbath in devout contemplation of the works of nature exhibited in the British Museum. "Within those walls," Tyndall argued, "we have, as it were, epochs disentombed—ages of divine energy illustrated."[32]

In 1856 Huxley declared his belief in the validity of natural theology:

> For man, looking from the heights of science into the surrounding universe, is as a traveller who has ascended the Brocken and sees, in the clouds, a vast image, dim and awful, and yet in its essential lineaments resembling himself. The numberless facts which illustrate this truth are familiar to all, through the works of Paley and the natural theologians, whose arguments may be summed up thus—that the structure of living beings is, in the main, such as would result from the benevolent operation, under the conditions of the physical world, of an intelligence similar in kind, however superior in degree, to our own.[33]

Throughout his life Huxley put forward strong expressions of his belief in a natural order, but he retreated from directly identifying God with that order after the seventies.

In 1859 he avowed that science was a divinely sanctioned activity that confirmed the order in both the physical and mental worlds. "The winning of every new law by reasoning from ascertained facts," Huxley declared, "the verification by the event, of every scientific prediction, is, if this world be governed by providential order, the direct testimony of that Providence to the sufficiency of the faculties with which man is endowed, to unravel, so far as is necessary for his welfare, the mysteries by which he is surrounded." Every law of nature which man had uncovered was to Huxley's mind "so many signs and wonders, whereby the Divine Governor signifies his approbation of the trust of poor and weak humanity, in the guide which he has given it."[34]

But by 1887, in "Scientific and Pseudo-Scientific Realism," Huxley's vision of order is shorn of its explicit godly trappings. The spiritual object of the scientist was the investigation of a universe that was like a "sort of kaleidoscope, in which, at every successive moment of time, a new arrangement of parts of exquisite beauty and symmetry would present itself; and each of them would show itself to be the logical consequence of the preceding arrangement, under the conditions which we call the laws of nature." Gazing upon this sight, a spectator might be filled "with that *Amor intellectualis Dei*, the beatific vision of the *vita contemplativa*, which some of the greatest thinkers of all ages, Aristotle, Aquinas, Spinoza, have regarded as the only conceivable eternal felicity." Huxley concluded his hymn to nature by stating matter of factly that "order is lord of all" (*SCT*, 74).

The pervasive existence of order in nature became the basis of a new natural theology formulated by the agnostics. The stress on the natural order was not in itself the novel element in agnostic natural theology. Indeed, it is this very emphasis on orderliness which places the agnostics within the tradition of natural theology. What the agnostics offered was new in the sense that it synthesized natural theology with modern science. Incorporated into this updated version of natural theology were the findings of evolutionary theory and the naturalism of the avant-garde within the new scientific elite.

Evolution and Natural Theology

Portraying the agnostics as the new natural theologians appears, at first glance, to be somewhat absurd in light of our general picture of the relationship between evolutionary theory and the design argument. It is ordinarily said that Darwin effectively undermined Paley's classical design argument for the existence of God by demonstrating that blind and gradual adaptation could produce those contrivances of nature traditionally pointed to as evidence for purposeful design. But although the agnostics acknowledged that Paley's version of natural theology was indeed demolished, nevertheless they believed that the theory of evolution reinforced, rather than discredited, the idea of order in nature. Like other laws of nature, it pointed to the operation of uniformity throughout the universe. The key to unraveling this seeming contradiction lies in the agnostic attitude toward Darwin's theory of natural selection.

The agnostics' support of scientific naturalism led them to take an interest in Darwin's *Origin of Species.* What they liked about the book was its attempt to present a theory of descent in purely natural terms. Their primary commitment was always to scientific naturalism and not to the survival of a particular theory of evolution. However, since Darwin had been attacked on theological grounds, they were willing to use the occasion of the publication of the *Origin* as a focus for their defense of a science purged of all references to metaphysical entities.

Huxley's "Darwinian Hypothesis," which appeared anonymously in the *Times*, on 26 December 1859, was one of those articles that secured for *The Origin of Species* a fair hearing in England. But a central point that Huxley insisted upon in this piece was the provisional quality of Darwin's theory of natural selection. Although immensely superior to other naturalistic explanations of organic descent, the truth or falsehood of Darwin's theory was impossible to determine without further information. Huxley recommended an "active doubt" as the state of mind proper to Darwin's hypothesis. "The combined investigations of another twenty years may," Huxley believed, "perhaps, enable naturalists to say whether the modifying causes and the selective power, which Mr. Darwin has satisfactorily shown to exist in Nature, are competent to produce all the effects he ascribes to them; or whether, on the other hand, he has been led to over-estimate the value of the principle of natural selection as greatly as Lamarck over-estimated his *vera causa* of modification by exercise." But almost twenty years later, in 1878, Huxley expressed his uncertainty about natural selection in his article "Evolution in Biology." Although Darwin, unlike Lamarck, had offered more evidence for accepting evolution, there was still some doubt in Huxley's mind as to the validity of natural selection. "How far 'natural selection' suffices for the production of species remains to be seen. Few can doubt that, if not the whole cause, it is a very important factor in that operation."[35]

In 1880 Huxley wrote to Darwin concerning an article that the former had published entitled "On the Coming of Age of the Origin of Species." Huxley explained to Darwin why he had not mentioned the theory of natural selection in an article that supposedly focused on the triumph of Darwinian theory:

> I hope you do not imagine because I had nothing to say about "Natural Selection," that I am at all weak of faith on that article. On the contrary, I live in hope that as palaeontologists work more and more . . . we shall arrive at a crushing accumulation of evidence in that direction also. But the first thing seems to me to be to drive the fact of evolution into people's heads; when that is once safe, the rest will come easy. (*LLTHH* 2:13)

Darwin was not so sure of Huxley's "faith," and he replied in a letter of 11 May 1880 that he agreed with Huxley's motive for not referring to natural selection. "But at the same time," he added, "it occurred to me that you might be giving it up, and that anyhow you could not safely allude to it without various 'provisos' too long to give in a lecture."[36] Darwin then launched into a discussion of various recent confirmations of natural selection, revealing his fear that Huxley was not a true believer. He was right to wonder, and even as late as 1892 Huxley still viewed natural selection as a hypothesis, for he said, "[that] the doctrine of natural selection presupposes evolution is quite true; but it is not true that evolution necessarily implies natural selection."[37]

Since their primary concern was to encourage the use of naturalistic explanations in science, the agnostics were loathe to put all their eggs in the Darwinian basket of natural selection.[38] The proper scientific attitude was to wait and see if the empirical evidence confirmed the validity of Darwin's hypothesis. The agnostic ambivalence toward natural selection, and the randomness it implied, accounts for what would otherwise be a mystifying tendency in their attitude toward evolutionary theory. Stephen maintained that evolution was an orderly process and that therefore no antitheistic conclusions could be drawn from the fact that it operated in the natural world. "It may be fairly urged," Stephen announced, "that a theory which tends to bring order out of chaos, and to reveal some general scheme working throughout all time and space, renders it more easy to maintain such rational theism as is now possible." Although Darwinism was incompatible with Paley in that God could no longer be conceived of as intervening directly to design each contrivance in nature, Stephen declared that the *Origin* did not rule out conceiving of God as the author of the laws of nature which generate the purposefulness in the natural world (*FP*, 99–100). Tyndall was also perfectly sincere when he reassured the Victorian reader of the congruence of evolution and Christian theism: "Fear not the Evolution hypothesis. . . . Under the fierce light of scientific enquiry, it is sure to be dissipated if it possess not a core of truth. Trust me, its existence as a hypothesis is quite compatible with the simultaneous existence of all those virtues to which the term 'Christian' has been applied. It does not solve—it does not profess to solve—the ultimate mystery of this universe" (*FS* 2:133). Both Tyndall and Stephen argued that evolution was not antitheistic since, in the final analysis, it pointed to order in the universe.

Like Stephen and Tyndall, Huxley saw no special antagonism between evolution and Christian theism. Huxley emphasized that evolution, although destroying forever the previous view of order, still provided for a broader teleological conception of nature:

> The teleology which supposes that the eye, such as we see it in man, or one of the higher vertebrata, was made with the precise structure it exhibits, for the purpose of enabling the animal which possesses it to see, has undoubtedly received its death-blow. Nevertheless, it is necessary to remember that there is a wider teleology which is not touched by the doctrine of Evolution, but is actually based upon the fundamental proposition of Evolution. This proposition is that the whole world, living and not living, is the result of the mutual interaction, according to definite laws, of the forces possessed by the molecules of which the primitive nebulosity of the universe was composed.[39]

In the pages of the *Fortnightly Review* Huxley stressed the theological affinities of Darwinism in a colorful fashion. Evolution, Huxley declared, was the scientific parallel of the Christian doctrine of Providence.[40]

The agnostics were among the first Victorians to recognize that the natural theology tradition could only survive if it were revised to harmonize with the new findings of evolutionary research. Their general approach shifted the emphasis from the particular to the universal. Purpose, order, and teleology were considered from the viewpoint of the whole of nature. Any specific cases that appeared to be instances of disorder could be dismissed in light of the whole. Even Clifford, who, unlike Huxley and Tyndall, had no reservations about accepting natural selection, saw nature as orderly and subordinated dysteleologies to larger ends. A number of important orthodox Christian theologians, discussed by James Moore in his chapter on "Christian Darwinism" in *The Post-Darwinian Controversies,* later came around to the agnostic way of thinking. They, too, learned to accommodate natural theology to evolutionary theory by offering a revised Paleyism based on a broader concept of order in nature (252–345).

Evolution and the Religion of Science

In addition to grounding a new vision of natural theology, evolutionary theory and science in general were often objects of religious reverence to earnest agnostics. In contrast to what Stephen saw as the vulgar, unpoetical spirit of Christianity, he put forward the tolerant, truly religious nature of evolutionary theory. The true evolutionist recognized the value of the religious instinct in human beings and admitted the importance of finding a means of embodying it in the future. Stephen hoped that evolutionary theory would provide a religion that would appeal both to emotion and to logic (intellect) and that would be expressed through a synthesis of sound philosophy and elegant poetry.[41] Clifford also saw in evolution the key to a new religion. He belonged to a circle of young men enthused with the potential of evolutionary theory, who looked to evolution for a new system of ethics which would combine the precision of the utilitarian with the poetical ideals of the transcendentalist (*LE* 1:33).

Francis Galton, referred to by his biographer, Karl Pearson, as a "religious agnostic," attempted to raise Darwinian doctrine to the status of a religious creed.[42] Galton first experienced doubts about orthodox Christianity during the mid forties, when visits to the Middle East exposed him to non-Christians whose conduct compared favorably with his own coreligionists. His admiration of the Mohammedans led him to be more tolerant and to take a broader view of the function of religion.[43] Darwin later supplied Galton with a positive faith. Galton believed that it was the individual's sacred duty to advance the evolutionary process for the sake of the race. "The direction of the emotions and desires towards the furtherance of human evolution," Galton declared, "recognized as rightly paramount over all objects of selfish desire, justly merits the name of a religion."[44] Since he saw eugenics as arising out of the ideas presented in the Sermon on the Mount, Galton looked forward to the day when the old faiths would remold themselves to new scientific ideas.[45]

Science in general could also serve as the basis of agnostic religion. Despite the separation of science and religion into two realms of authority, agnostics often saw religious feeling and scientific fact as pointing to one and the same God. Huxley argued that only a new religion centered on a theology generated by science would be acceptable. He hoped that science would eventually purge religion, as well as Spencerian agnosticism, of all false theology. But, like the other agnostics, his natural theology led him to treat science as the source of a new religion. In 1859 he observed: "we are on the eve of a new Reformation and if I have a wish to live thirty years, it is that I may see the God of Science on the necks of her enemies. But the new religion will not be a worship of the intellect alone." Stephen's discussion of the religion of the future in his essay "The Religion of All Sensible Men" pointed to science as "the fixed fulcrum, an unassailable nucleus of definite belief, round which all other beliefs must crystallise." Before Clifford died he sketched out plans for recasting his lectures and writings into a book to be called *The Creed of Science*. When Tyndall gave his last lecture at Queenwood in 1848 before going off to Marburg, he told his students that he wanted to study the physical sciences not only to know about natural things, but also to come closer to God. "What are sun, stars, science, chemistry, geology, mathematics," Tyndall believed, "but pages of a book whose author is God! I want to know the meaning of this book, to penetrate the spirit of this author and if I fail then are my scientific attainments apple rinds without a core."[46]

Although the agnostics were content to treat science as a religion since it was the study of divine natural law, they would have reacted in a hostile manner to the suggestion that the primary principles upon which science was based were mere articles of faith. However, by severely limiting knowledge, the agnostics inadvertently created problems for themselves in their attempt to justify the validity of scientific principles.

The Agnostics and A Priori Knowledge

In 1882 Henry Sidgwick submitted the agnostic theory of knowledge to a devastating critique. "I find myself unable," he concluded, "with all the aid of the eminent thinkers who have recently maintained some form or other of Empiricism—to work out a coherent theory of the criteria of knowledge of an Empirical basis."[47] As Sidgwick knew only too well, his inability stemmed from the weaknesses in agnostic assumptions about nature and the ability of human beings to possess certain knowledge of nature. The sceptical element in the Victorian agnostics' thought made it difficult for them to acknowledge the reality and validity of all first principles, whether they be religious, philosophical, or even scientific. Essentially sceptical and idealistic, agnosticism was an uneasy ally of scientific naturalism, which was basically dogmatic and materialistic. Like Spencer, the agnostics at times admitted that ultimate scientific ideas contained unresolvable contradictions.[48] But more often they clung to a determined faith in those axioms grounding science despite their inability to demonstrate their legitimacy with certainty. Besides Sidgwick, critics of scientific naturalism like James Ward, George John Romanes, Samuel Butler, and Arthur James Balfour pointed out that the agnostics were representatives of a rationalist orthodoxy as rigorous and restrictive as its theological counterpart.[49]

W. H. Mallock, another opponent of agnosticism, used a slightly different approach to illustrate the shortcomings of scientific naturalism. In his novel *The New Republic* (1877) Mallock presented caricatures of Huxley, Tyndall, and Clifford (the models for the characters of Storks, Stockton, and Saunders). Throughout the book Clifford is portrayed as a young hothead who constantly embarrasses Huxley and Tyndall by putting forward their position too starkly. " 'And the worst of it is,' said Mr. Storks [the Huxley character], 'that these young men really get hold of a fact or two, and then push them on to their own coarse and insane conclusions—which have, I admit, to the vulgar eye, the look of being obvious' " (183). Of all the agnostics, Clifford was the most aware of the epistemological difficulties generated by his agnosticism. He recognized that from his position the first principles of science could only be contingent, and therefore subject to revision.

The agnostics believed that the only valid type of knowledge possible was obtained through science or the study of experience. Stephen represented agnostic thinking well when he stated that "when we see daily with more clearness that all intellectual progress involves a systematic interpretation of experience and a resolute exclusion of all imaginary *a priori* data, it is desirable that we should look in the direction in which alone experience can enlighten us, and accept realities in exchange for dreams" (*AA*, 84). This repudiation of intuitionism was part of the agnostics' debt to Mill, Spencer, and Benthamite empiricism.

However, in sharing Mill's assumption that there is no a priori knowledge, the agnostics also ended up forsaking Kant's conception of the limits of knowledge. They embraced what Kant would have described as transcendental realism. Mathematics and geometry are not, according to the agnostics, bodies of synthetic a priori knowledge.[50] Furthermore, the agnostics denied that space and time are a priori forms of intuition.[51] Drawing upon both Mill's association psychology and Spencer's evolutionary explanation of the so-called a priori elements of the mind, the agnostics argued that the necessity apparently contained by the truths of mathematics and geometry, the forms of intuition, and the categories of the understanding were merely the result of a feeling of strong association built up through eons of experience.[52] Like Tyndall, Stephen, and Clifford, Huxley insisted that "in his 'Principles of Psychology,' Mr. Herbert Spencer appears to me to have brought out the essential truth which underlies Kant's doctrine in a far clearer manner than any one else."[53]

Stephen admitted that his rejection of a priori knowledge left science "empirically and radically uncertain"; however, he asserted that this confession had no effect on everyday life. Although we cannot ground science, it does not, to Stephen, make any difference, because we assume it in all our actions. This attitude does not stop Stephen from maintaining that scientific doctrines are established truths, because they are verified and proved by their predictive powers. Clifford was even more radical than Stephen. He asserted that the postulates of all the exact sciences "are not, as too often assumed, necessary and universal truths; they are merely axioms based on our experience of a certain limited region." It was, in part, Clifford's work in non-Euclidean geometry which helped him to face the implications of his standpoint.[54]

Clifford was deeply influenced by G.F.B. Riemann's non-Euclidean geometry, which, to him, indicated that geometrical truth is derived from experience.[55] Clifford had learned from Riemann that geometry is a formal exercise in logic when considered as a pure science of ideal space, but when applied to actual space, it is a physical science and is therefore subject to verification.[56] The geometrical calculations of the past are "practically exact" for the finite things that we presently deal with, "yet the truth of them for very much larger things, or very much smaller things, or parts of space which are at present beyond our reach, is a matter to be decided by experiment, when its powers are considerably increased." Clifford maintained that we should assume the practical universality of geometry and mathematics (knowing it is not theoretically universal) as it "pays us to assume it" (*LE* 1:137–38). Utility is the criterion of verification.

Clifford expressly attacked Kant's use of geometry to prove the truth of his "Transcendental Aesthetic." He recognized that Kant's position was impregnable once one admitted that geometrical knowledge was exact, universal, and necessary. "To any one who admitted the necessity," Clifford declared, "the argument was even stronger; for it was clear that no experience could make any approach to supply knowledge of this quality" (*LE* 1:269). In Kant's dictum "given knowledge, how is it possible?" Clifford made Kant's transcendental deduction of space and time stand or fall with Euclidean geometry ("given knowledge"), and it was Clifford's adherence to non-Euclidean geometry which led him to dismiss Kant hastily.

Clifford claimed that space may possibly be either finite or infinite, and that these conceptions are perfectly conceivable. Kant's antinomy of space, to Clifford, arose from "the assumption of theoretical exactness in the laws of geometry" (i.e., Kant's assumption of the validity of Euclidean geometry) (*LE* 1:154). Clifford even went as far as asserting in one of his mathematical papers that space is curved, and he has been considered in this respect to be one of Einstein's forerunners.[57] But though Einstein rejected Euclidean geometry, he did not undermine Kant's notion of space and time as a priori forms of intuition. Einstein and Kant disagree on the structure of space. Yet relativity theory implicitly accepts that the structure of our perceptual organization is given a priori and that space is not a thing but one of the forms through which we organize our perception of things.[58] Unlike Clifford, Einstein resolves the antinomy of space without destroying Kant's main position that there is an interrelationship between human beings and nature.

Clifford met Kant's epistemological question concerning the possibility of knowledge head on. He realized that the Kantian dilemma allowed a choice only between synthetic a priori judgments or the rejection of all universal statements. "Either I have some source of knowledge other than experience," Clifford asserted, "and I must admit the existence of *a priori* truths, independent of experience; or I cannot know that any universal statement is true. Now the doctrine of evolution itself forbids me to admit any transcendental source of knowledge; so that I am driven to conclude in regard to every apparently universal statement, either that it is not really universal, but a particular statement about my nervous system, about my apparatus of thought; or that I do not know that it is true." Since Clifford did not admit any a priori synthetic knowledge, he realized that he must consistently affirm that no universal statements are possible. Any supposed universal statement is either "a particular statement about my nervous system" (e.g., the categories, space, and time as explained by Spencer) or not capable of being known to be true (e.g., mathematics and geometry). All knowledge, Clifford consistently admitted, is "of the nature of inference, and not of absolute certainty." Clifford's position, as Richards argues, was no doubt useful for the scientific naturalists' attempt to undermine the Christian intuitionalists' claim to possess absolute truth in geometry as well as in nonmathematical areas. But Clifford's radical uncertainty was also a potential threat to the pretensions of science to possess an intellectual superiority over orthodox Christian doctrine based on uncertain faith.[59]

The Dogma of Uniformity in Nature

The agnostics considered the idea of uniformity in nature, the concept of cause and effect, and the notion of an external, natural world to be axioms necessary to science. But by rejecting a priori knowledge and Kant's transcendental idealism, they were forced to justify these principles while remaining on an empirical level. Kant argued that it was impossible to legitimize universal concepts such as uniformity in nature or cause or to derive a ground of necessity from the transcendental realist/empirical idealist perspective. He also learned from Hume that the existence of an external world of physical things could not be demonstrated if one started from empiricist assumptions. The agnostics struggled in vain to overcome the inadequacies of their epistemological position as they came to grips with the three articles of faith which comprised their holy trinity.

The agnostics asserted that the idea of the uniformity of nature was universal, necessary, and absolutely vital to science. "To deny it," Stephen insisted, "is to fall into absolute scepticism, for it is to cut out the very nerve of proof in every proposition drawn from experience." Huxley echoed Stephen's thoughts: "The fundamental axiom of scientific thought is that there is not, never has been, never will be, any disorder in nature. The admission of the occurrence of any event which was not the logical consequence of the immediately antecedent events, according to these definite, ascertained, or unascertained rules which we call the 'laws of nature,' would be an act of self-destruction on the part of science." Tyndall seconded Huxley's point. "Has this uniformity of nature ever been broken? The reply is: 'Not to the knowledge of science.'" In 1866, Tyndall jokingly chided Huxley for his lack of zeal in defending the absolute validity of the principle of uniformity. If Tyndall were to say that he could by a word cause a stone to fall upward, Huxley declared that he would feel bound to suspend his judgment until the matter could be investigated. Tyndall thanked Huxley for the compliment, but felt that were the situation reversed his judgment regarding his friend would be best expressed in the line "'Oh what a noble mind is here o'erthrown!'"[60]

Clifford also gave the conception of the uniformity of nature an important role in his philosophy of science. Clifford distinguished between mere technical thought (which enables us to deal with circumstances encountered previously) and scientific thought (which allows us to deal with different circumstances that we have never met with before). The key to science's ability to cope with new situations is the uniformity of nature:

> The aim of scientific thought, then, is to apply past experience to new circumstances; the instrument is an observed uniformity in the course of events. By the use of this instrument it gives us information transcending our experience, it enables us to infer things that we have not seen from things that we have seen; and the evidence for the truth of that information depends on our supposing that the uniformity holds good beyond our experience. (*LE* 1:131)

All of the agnostics conceived of the uniformity of nature as a grounding principle which, if denied, destroyed science.

Kant had guaranteed the a priori validity of the idea of the uniformity of nature by formulating his notion of the categories. It was the categories which made experience possible through their organization of the manifold of intuition, and this process made nature uniform. However, the agnostics had rejected all a priori synthetic knowledge, and hence could not justify the conception of the uniformity of nature in this way. Their various attempts at justification were doomed to failure.

Stephen admitted that although the postulate of the uniformity of nature is continually verified by experience, this fact was not sufficient proof of its validity as a certain, a priori principle. However, he maintained that this was proof enough for a belief assumed for the sake of action. "Though such a process," Stephen declared, "however far it is continued, can never reach a certainty, it may amply justify a postulate; that is, a belief assumed for purposes of action." Another argument that Stephen put forward was that we have no choice but to assume the uniformity of nature. "It seems to me that in any case the formula is not so much a postulate, universal or otherwise," Stephen declared, "as a statement of the process which constitutes all reasoning about facts. The alternative is not assuming some other postulate, but ceasing to think about reality." Stephen never stops to think what is signified by the fact that we must assume this postulate or we cannot reason. Stephen's view was shared by Huxley, who stated that the "general constancy of the order of Nature" is one of a group of hypothetical assumptions "which cannot be proved, or known with that highest degree of certainty which is given by immediate consciousness; but which, nevertheless, are of the highest practical value, inasmuch as the conclusions logically drawn from them are always verified by experience."[61]

Clifford, too, talked of the pragmatic necessity of assuming the conception of the uniformity of nature. Scientific inference, Clifford maintained, depends on the assumption of the uniformity of nature. But what does this rest on, Clifford asks? "We cannot infer that which is the ground of all inference" he replies. We do know that "nature is selecting for survival those individuals and races who act as if she were uniform" (*LE* 1:293–94). Clifford's reasoning has a circular quality. The assumption of uniformity is a presupposition of scientific inference, but the advance of scientific theory justifies the principle of uniformity.[62] Clifford ends up attributing to the principle of the uniformity of nature the same status he gave to mathematics and geometry. We cannot believe that "nature is absolutely and universally uniform," but we may assume that nature is "practically uniform so far as we are concerned" (*LE* 2:210).

The Doctrine of Cause

A similar problem exists for the agnostics when it comes to a notion of cause and effect. Tyndall, Stephen, and Huxley all asserted that a conception of cause is crucial for science, but they were unwilling to admit that it is a priori. For example, Huxley asserted that physical science starts from certain postulates, one of which "is the universality of the law of causation; that nothing happens without a cause (that is, a necessary precedent condition), and that the state of the physical universe, at any given moment, is the consequence of its state at any preceding moment." In another essay, Huxley stressed that modern science had made it inconceivable "that chance should have any place in the universe, or that events should depend upon any but the natural universe, or that events should depend upon any but the natural sequence of cause and effect." Huxley used the notion of cause in his attack on miracles, and he clearly emphasized the necessity implicit in cause and effect. "When repeated and minute examination never reveals a break in the chain of causes and effects," Huxley asserted, "and the whole edifice of practical life is built upon our faith in its continuity; the belief, that that chain has never been broken and will never be broken, becomes one of the strongest and most justifiable of human convictions."[63]

Stephen backed Huxley up on this issue and perceived an intimate connection between evolution, science, and a notion of cause. "The doctrine of evolution," Stephen affirmed, "is the uncompromising application to all phenomena of history and thought of a genuine belief in causation, or of an expulsion of the arbitrary." Stephen viewed the notions of the uniformity of nature and the universality of causality as phrases signifying the same thing. To him they were not propositions to be tested for their truth or falsehood, rather they were assumptions that must ground all reasoning. The alternative to accepting them was "to cease to think." Likewise, Tyndall was in full agreement with Huxley and Stephen that there was no "element of caprice" in nature. In fact, when humanity recognized this principle ages ago, it made a tremendous leap forward. "The notion of spontaneity," Tyndall maintained, "by which in his ruder state [man] accounted for natural events, is abandoned; the idea that nature is an aggregate of independent parts also disappears, as the connection and mutual dependence of physical powers become more and more manifest: until he is finally led to regard Nature as an organic whole—as a body each of whose members sympathises with the rest, changing, it is true, from age to age, but changing without break of continuity in the relation of cause and effect."[64]

But, despite Huxley, Stephen, and Tyndall's insistence that there is a chain of cause and effect in nature, they refused to give this knowledge a priori status. Huxley, for example, after reading Hume and Mill, accepted their associationist explanation of the deceptive feeling of necessity accompanied by cause and effect.[65] In addition to rejecting cause as a priori, he also recognized that the universality of the law of causation "cannot be proved by any amount of experience" (*EE*, 121). Hence, the validity of this postulate was, as Huxley admitted, undemonstrable.

In 1866 Huxley remarked that since observation and experiment could confirm that nature operates in a regular fashion, a belief in the uniformity of nature was justification "not by faith, but by verification." However, in later life Huxley was willing to stake the validity of important scientific axioms on faith. "The one act of faith in the convert to science," Huxley wrote, "is the confession of the universality of order and of the absolute validity in all times and under all circumstances, of the law of causation. This confession is an act of faith, because by the nature of the case, the truth of such propositions is not susceptible of proof." Huxley was forced to resort to an argument that was based on an unscientific principle. He affirmed that one must have "faith" that cause is universal, and that such "faith" was reasonable because it was confirmed by experience and because it operated as the foundation of all action. Once again, the way out was an appeal to experience and pragmatism. This attempt at justifying the notion of cause left Huxley open to the charge of hypocrisy, for it was difficult to perceive how holding to a scientific axiom without adequate evidence differed from the faith in God expressed by orthodox Christians which Huxley frequently attacked.[66]

Clifford was an important exception to the other agnostics' view of cause. Where Huxley, Stephen, and Tyndall attempted to hold on to a notion of cause in nature while simultaneously rejecting its a priori quality, Clifford seemed to recognize the weakness of this position. He refused to emphasize cause and effect as much as the other agnostics. He believed that the term *cause* has many different meanings, and that when one particular sense of the word helps us to understand a sequence of events, we try to apply it "as a simile to all other events." When we meet a case where the simile does not apply, we refuse to confess that it was only a simile and to assert that the cause of the event is a mystery. "When we say then that every effect has a cause," Clifford contended, "we mean that every event is connected with something in a way that might make somebody call that the cause of it. But I, at least, have never yet seen any single meaning of the word that could be fairly applied to the *whole* order of nature" (*LE* 1:150–51). Clifford's de-emphasis of cause and effect is a reflection of his allegiance to the new statistical view of nature, and this aspect of his thought sometimes separated him from the other agnostics when it came to dealing with scientific theory and the philosophy of science.

Clifford and the Radical Nature of Probabilistic Law

Clifford's dissent from the agnostic doctrine of cause, and his willingness to recognize that the empiricist position made it impossible to hold with justification the existence of universality in scientific theory, stemmed in part from his acceptance of probabilistic methods. Recent developments in science, particularly atomic theory, raised a whole series of epistemological issues for Clifford which ultimately lent to his agnosticism a more radical complexion. During the last half of the nineteenth century, a revolution was occurring in scientific thought which was setting the stage for twentieth-century probabilistic theories. The old Newtonian notion of science, which demanded a mechanistic, deterministic view of nature, was being replaced by the rise of a statistical approach to science which culminated in quantum physics. Until the middle of the nineteenth century, scientists were only willing to accept deterministic laws as valid. Statistical laws were first used in the social sciences by Quetelet, Buckle, Marx, and Engels. But by 1859, statistical laws appeared in physics with Maxwell's lecture on kinetic theory, and in biology when Darwin published the *Origin of Species*.[67]

The molecular-kinetic theory of gases was the first statistical theory in physics, and only in the second half of the nineteenth century was it explicitly formulated by scientists such as Krönig, Clausius, and Maxwell. Maxwell's chief contribution was the distinction between two methods for studying natural phenomena. The "dynamical method" was to be applied when it was possible for the scientist to observe the motions of individual particles, while the "statistical method" was more appropriate when one was dealing with a number of particles large enough to make it impossible to follow each one separately.[68] The implications of this idea were staggering. Maxwell was arguing that, because of the innumerable number of particles in a gas, and because of our inability epistemologically to grasp these entities individually, scientists must deal with microphenomena (i.e., the atomic world) differently from the way they study macrophenomena (i.e., the natural world and the world of astronomy). Scientists must give up the old concise, precise, and mechanistic Newtonian analysis based on cause and effect relationships when attempting to investigate the structure of the atomic world. Maxwell therefore set out to develop new techniques to describe the atomic universe. Statistics could deal with aggregates, but, as Maxwell argued, the statistical method yielded averages and unvarying regularities, and not absolute certainty in every individual case.[69]

Darwin's theory of natural selection was also a statistical or probabilistic law, although Darwin did not describe it in these words. Natural selection dealt with species on the level of whole populations and not individual organisms. It was precisely the application of statistical methods in Darwin's theory which led some biologists to reject it, because to them natural selection introduced blind chance into science (i.e., it was not Newtonian).[70]

Both Maxwell and Darwin believed that, for the time being, scientists must be content with a partial description of natural phenomena. Although their scientific laws were probabilistic, they conceived of nature as subject to deterministic laws.[71] Developed in the early twentieth century, quantum theory made the next step by considering both nature and science to be probabilistic. The abandoning of "dynamical theory" implied to scientists such as Heisenberg and Bohr that a purely objective and mechanical description of nature was no longer possible. People must participate in organizing nature in order to make sense of it.[72] Both of these scientists conceived of nature as being related to human beings, and they held to a form of Kant's transcendental idealism. Therefore, the development of probabilistic methods and quantum theory made it increasingly difficult for scientists to hold to transcendental realism as the nineteenth century drew to a close and the twentieth century progressed. It also challenged the validity of a deterministic view of nature.

Clifford began formulating his epistemology at the very start of this revolution in science. The problem for him was to decide whether he would accept atomic theory as valid, and, if so, how to come to grips with the epistemological implications of probabilistic law. This decision was no easy matter in the sixties and seventies. Atomic theory seemed to work in the physical sciences, but there was no mechanical means by which scientists could prove its legitimacy, because instruments had not yet been invented which allowed one to perceive the atomic world. As a result, some scientists (particularly chemists) banished atomic theory to the realm of metaphysics up to the first decade of the twentieth century.[73]

Huxley was far too empirically minded to swallow atomic theory whole. He wrote to Kingsley that "I don't know that atoms are anything but pure myths" (*LLTHH* 1:261). When Tyndall professed to see atoms visually in his mind's eye at an X-Club meeting, but was unable to describe what he saw, Huxley replied "Ah, now I see myself; in the beginning was the Atom, and the atom *is* without form and void, and darkness *sits* on the face of the Atom!"[74] Tyndall believed that scientists who shared Huxley's refusal to acknowledge the reality of atoms were unwisely cautious.[75] But even Tyndall, along with Stephen and Huxley, believed that science would eventually reduce the problems of atomic physics to questions of mechanics. The scientific imagination, Tyndall declared, could apply to the atomic world "reasoning as stringent as that applied by the mechanician to the motions and collisions of sensible masses."[76] Clifford recognized the implications of accepting probabilistic laws such as natural selection and atomic theory. Because he was the youngest of the agnostics, he had grown up under the influence of Darwin's views and was able to work out a truly evolutionary worldview. Like scientists such as Boltzmann, Clifford was aware that the indeterministic quality of probabilistic theory raised serious questions about the validity of Newtonianism.[77] Like Peirce and the other American pragmatists, Clifford saw the revolutionary implications of evolution for science. Whereas the other agnostics perceived evolution as merely reinforcing the old mechanistic, determinist approach to science, Clifford linked Darwinian theory to probabilistic law.[78]

Clifford was familiar with Maxwell's kinetic theory of gases, and he asserted that "what is called the 'atomic theory' . . . is no longer in the position of a theory, but that such of the facts as I have just explained to you are really things which are definitely known and which are no longer suppositions." Clifford was willing to assert the existence of particles not perceived by the senses. His argument for the validity of atomic theory was basically pragmatic. He credited Maxwell and Clausius with demonstrating not only that the molecular theory of matter explains the facts, but "also that no other will." This theory, then, was not a guess but an "organised account of the facts, such that from it you may deduce results which are applicable to further experiments, the like of which have not yet been made" (*LE* 1:163, 195–97).

Clifford fully realized that atomic theory was probabilistic and could not yield certainty. "But the law is one of statistics," Clifford affirmed, "its accuracy depends on the enormous numbers involved; and so, from the nature of the case, its exactness cannot be theoretical or absolute" (*LE* 1:139). Clifford was also conscious of the fact that the statistical view of nature was slowly undermining the old Newtonian approach to science, which depended on theoretical exactness and mathematical law. Some philosophers, Clifford observed, tried to avoid this conclusion by using epistemological subterfuge:

> As the discoveries of Galileo, Kepler, Newton, Dalton, Cavendish, Gauss, displayed ever new phenomena following mathematical laws, the theoretical exactness of the physical universe was taken for granted. Now, when people are hopelessly ignorant of a thing, they quarrel about the source of their knowledge. Accordingly, many maintained that we know these exact laws by intuition. These said always one true thing, that we did not know them from experience. Others said that they were really given in the facts, and adopted ingenious ways of hiding the gulf between the two. (*LE* 1:140)

Clifford's refusal to base his notion of science on a Newtonian determinism is closely related to his adherence to probabilistic methods. The rise of statistical reasoning, even prior to quantum physics, was one of the factors leading to the decline and fall of causality in scientific explanation.[79] The result for Clifford is a radical uncertainty in science which he cheerfully accepts.

Clifford asserted that the uniformity of nature "is not fixed and made once for all, but is a changing and growing thing, becoming more definite as we go on" (*LE* 2:138). He did not conceive of the uniformity of nature as being ultimate truth. It took on this deceptive air as only those races who acted as if it were absolute truth have survived. The status of the uniformity of nature, as well as all the laws of nature (for they are based upon the assumption of the uniformity of nature) is uncertain. The elevation of natural laws to the rank of scientific truth is completely random, because they are a product of chance and natural selection. The ability of human beings to discover absolute truth is limited by two factors, according to Clifford. First, nature itself is evolving, implying the evolution of the laws governing it. Second, human beings, while they attempt to understand an evolving natural world, are themselves subject to the process of evolution. Both the observers and the observed are in a state of flux. Evolution, to Clifford, has shown that we are part of nature both physically and intellectually. We are so much an integrated part of nature that we can never distance ourselves from the natural world in order to ground science. We are so involved in the random process of evolution that we can never have objective knowledge of nature.

Unlike the other agnostics, Clifford reached intellectual maturity in the sixties, in the midst of controversies surrounding the *Origin of Species*. For Clifford, Darwinian theory raised the issue of the status of science in an evolutionary worldview. The absolute quality of religion suffered erosion when historians and anthropologists began to study religious ideas and institutions as phenomena subject to evolutionary development. Darwin, agonizing over the problem of God's existence, doubted that the human mind, "developed from a mind as low as that possessed by the lowest animals," could be trusted "when it draws such grand conclusions."[80] Clifford asked the same question, but in regard to science. He wondered whether the human mind, a product of evolution, was reliable in its scientific as well as religious conclusions.

In perceiving that even the laws of nature evolve, Clifford was taking Darwin's concept of natural selection as seriously as possible. The other agnostics refused to take this step because it seemed to them that it destroyed the validity of science. In accepting the notion of evolution, Huxley, Tyndall, and Stephen did not feel it was necessary to transform radically their view of the philosophy of science. But for Clifford, in order to embrace natural selection, one was committed to a world of uncertainty where science would never supply precise and necessary knowledge of a constantly evolving natural world.

The Tenet of an External World

In 1873 Tyndall wrote to Huxley that during a friendly visit "Spencer once deprived me of a night's sleep by asking me just as I was retiring to bed whether I believed in the externality of matter."[81] Apparently this same incident left a deep impression on Spencer as well, for in 1894 he recalled the late-night conversation with Tyndall on the existence of matter and remarked "that persistence in this kind of thing was out of the question, and I had to abridge my stay."[82] In Spencer's version of the story it was Tyndall who was to blame for posing the question over which both men lost sleep.

It was no wonder that the issue of the existence of an external, objective natural world troubled the agnostics. Huxley asserted that science required the existence of an external world of nature as an axiom: "All physical science starts from certain postulates. One of them is the objective existence of a material world. It is assumed that the phenomena which are comprehended under this name have a 'substratum' of extended, impenetrable, mobile substance, which exhibits the quality known as inertia, and is termed matter" (*MR*, 60). But, as Henry Sidgwick and James Ward pointed out, the agnostics confronted problems when they attempted to demonstrate the validity of immediate cognition. The sceptical and idealistic tendencies of agnosticism "costs Naturalism, as it turns out, its entire philosophical existence," Ward maintained. "In order to be free of 'metaphysical quagmires' such as the ideas of substance and cause, it is led to reject the reality not only of mind, but even of matter; and in this state of ideophobia must collapse, for lack of the very ideas it dreads."[83]

The agnostics were presented with a dilemma when formulating their theory of knowledge. If they accepted Mansel's agnosticism in the realm of religious knowledge, did this standpoint necessarily entail an extension of agnosticism into the scientific sphere? Did a religious agnosticism demand a scientific empiricism that was ultimately self-destructive when it dismissed both God and an external world as unknown, transcendent entities? The agnostics, like Mansel, tended to set the limits of human knowledge at a point that did not allow them to take the stand Kant insisted upon as the only approach toward a justification of science.

The agnostics were not attracted to previous attempts by empiricists to defend the tenet of an external world. In *An Examination of Sir William Hamilton's Philosophy*, Mill had put forward a theory drawing on empiricist principles in order to provide a suitable alternative to Mansel and Hamilton's natural realism. Basically, Mill asserted that the belief in an external world was an acquired product. We weave our past experiences of sensation into a group of "permanent possibilities of sensation" by the law of association. The intuitionist approach to natural realism only seems plausible because we forget the process of association and conceive of the "permanent possibilities of sensation" as an external, independent nature (177–78, 183).

The agnostic reaction to Mill's theory is encapsulated in an amusing story told by Tyndall about his trip to Edinburgh in 1866 with Carlyle and Huxley. After Carlyle's successful speech as the new rector of the university, Carlyle, Tyndall, and Huxley attended a number of banquets, one of which celebrated the occasion with a round of songs and ditties. Lord Neaves (1800–1876), a Scottish judge, regular contributor to *Blackwood's Magazine*, and a satirist of some repute, taught the assembled company a song lampooning Mill's theory of knowledge, to the delight of Carlyle and Tyndall. The whole table joined in the refrain, "Stuart Mill on Mind and Matter," led by Carlyle, who waved his knife in the air as if it were a conductor's baton. To illustrate the absurdity of Mill's attempt to demonstrate the nonexistence of mind and matter, Neaves included a verse that ended "But had I skill, like Stuart Mill / His own position, I could shatter: / The weight of Mill, I count as Nil / If Mill has neither Mind nor Matter."[84] In Tyndall's estimate Neaves's song was "excellent."[85] Both Tyndall and Huxley witnessed the Scottish reaction to Mill's critique of their hero Hamilton, and both sided with Carlyle and the Scots.[86] Stephen, Huxley, Tyndall, and Clifford did not feel that Mill's attempt to overcome idealism was successful, nor did they draw upon the concept of the permanent possibilities of sensation to help them justify their belief in an external world. Stephen even pointed to the dangerous idealistic tendencies in Mill's theory.[87]

Although the agnostics rejected Mill's justification for the belief in an external, natural world, they were no more successful than he, primarily because they shared with Mill an adherence to Lockean sensationalism. The agnostics justified their belief in an external world in a variety of ways. Sometimes they claimed that we have a consciousness of an external world transmitted through the senses (though perhaps not an immediate consciousness). Other times they argued that, though the postulate of the objective existence of a material world (like all the other postulates of science) is unprovable, it is verified whenever tested by experience.[88] Huxley, Clifford, and Tyndall believed they could sidestep Hume's critique of sensationalism by relying on new developments in the study of physiology. Huxley was highly critical of the a priori metaphysicians, who attempted to base a theory of knowledge without taking physiology into account. "If the origin of the contents of the mind is truly a philosophical problem," Huxley declared, "then the philosopher who attempts to deal with the problem without acquainting himself with the physiology of sensation, has no more intelligent concept of his business than the physiologist, who thinks he can discuss locomotion, without an acquaintance with the principles of mechanics."[89]

However, in emphasizing physiology, Clifford, Tyndall, and Huxley were committed to a form of transcendental realism, for they conceived of the human nervous system as a factor that divided human beings and the external world. Tyndall certified that "between the mind of man and the outer world are interposed the nerves of the human body, which translate, or enable the mind to translate, the impressions of that world into facts of consciousness and thought." Huxley made the next logical step by asserting that since the nervous system is between the individual and the external world, therefore "no similarity exists, nor indeed is conceivable, between the cause of the sensation and the sensation." In holding to the transcendental realist stance, the agnostics were forced to conceive of the relation between nature (sensation) and the thing-in-itself (the external world and the cause of sensation) in causal terms, and hence were reduced to a reliance on inference in their justification of the existence of an external world. The agnostics' stress on physiology led straight to empirical idealism.[90]

There were times when the agnostics would admit publicly that their position, relying as it did on inferring from cause to effect to prove the existence of an external world, raised doubts as to whether the cause in question is in us or outside us. Tyndall declared that "all we hear, and see, and touch, and taste, and smell, are, it would be urged, mere variations of our own condition, beyond which, even to the extent of a hair's breadth, we cannot go. That anything answering to our impressions exists outside of ourselves is not a *fact*, but an *inference*, to which all validity would be denied by an idealist like Berkeley, or by a sceptic like Hume."[91] The agnostics argued that human beings can only directly know their own states of consciousness and sensations, and that, therefore, to consider sensations to be properties of external objects existing outside the mind would be to go beyond the limits of knowledge.[92] Huxley even sided with Berkeley against Locke on the issue of primary qualities, because the agnostic affirmed that they are not independent, self-existent entities.[93] Clifford also supported Berkeley on our inability to move beyond our own world of perceptions. "The physical universe which I see, and feel, and infer," Clifford affirmed, "is just my dream and nothing else; that which you see is your dream; only it so happens that all our dreams agree in many respects. This doctrine of Berkeley's has now been so far confirmed by the physiology of the senses that it is no longer a metaphysical speculation, but a scientifically established fact" (*LE* 2:142).

In order to demonstrate that life was not merely a dream, Clifford developed a very complex notion that he referred to as "mindstuff." He asserted the existence of absolute, independent entities which were part of all minds, and said that these things in themselves underlie nature and guarantee the reality of consciousness. Yet, even if Clifford were correct, his theory merely demonstrates that other minds are real and conscious. He fails to validate the belief in an external world of nature.[94]

Berkeleian agnosticism pictured nature as that which is produced by sensation and consciousness—it is an internal, solipsistic world. This notion of nature, for Kant, leaves science in a delicate position. Although we have knowledge of our minds, we can have no objective, universal knowledge (i.e., no knowledge that is proven to be shared by all). The concept of an objective, universal nature therefore disappears, for we are each locked up into our own world of dreams.

The agnostics claimed that, in comparison to orthodox theologians, they were more open to change, more aware of the status of scientific axioms, and more willing to subject their assumptions to scrutiny. "A law of nature," Huxley declared, "in the scientific sense, is the product of a mental operation upon the facts of nature which come under our observation, and has no more existence outside the mind than colour has" (*SCT*, 76). Huxley argued that law, force, and ether were merely useful symbols, rather than adequate expressions of reality. It was the theologians who, by conceiving of symbols as real existences, forgot that they were worshiping idols fabricated by human hands (*SHT*, 372). However, from the agnostic viewpoint, the articles of their holy trinity could not be mere symbols, or science would lose its claim to superiority over Christian orthodoxy. A law of nature as formulated according to the present state of scientific progress may be a product of human thought, however, the agnostics faithfully believed that embedded in nature there were laws that grounded the natural order.

The agnostics were too busy worshiping their holy trinity to recognize their own handiwork. Rather than compare agnosticism to a new form of idolatry, James Ward likened scientific naturalism to a humanly created monster that was running amok and out of control. "The man of science," Ward declared, "like Frankenstein, has conjured up this monster; and now pretending to have made *him* it pronounces him to be impotent, and the Nature it presents, to be the only One and All that he can ever know."[95] But if the possibility of knowledge was destroyed by weaknesses in agnostic assumptions about nature, the "man of science" could not even say that the "One and All" was ever known.

Conclusion

THE TRAGEDY OF AGNOSTICISM

> *It was absolutely necessary that the soul should be cut away. Religious belief, in the form in which we had known it, had to be abandoned. . . . For two hundred years we had sawed and sawed and sawed at the branch we were sitting on. And in the end, much more suddenly than anyone had foreseen, our efforts were rewarded, and down we came. But unfortunately there had been a little mistake. The thing at the bottom was not a bed of roses after all, it was a cesspool full of barbed wire. . . . So it appears that amputation of the soul* isn't *just a simple surgical job, like having your appendix out. The wound has a tendency to go septic.*
>
> GEORGE ORWELL (1940)

By the early 1890s agnosticism as a distinct movement had begun to wane.[1] Clifford had been buried over a decade earlier, in 1879. Those that remained, Spencer, Tyndall, Huxey, and Stephen, were tired, old, and found themselves living in a world in which their cause was increasingly irrelevant.[2] The once-energetic Tyndall was plagued by sleeplessness, and in 1889 wrote one of many entries in his journal concerning the misery he endured. "A poor night," he recorded, "the nights generally are poor now. It is a tragic life!"[3] The X-Club ceased to meet after 1892 because its members were either dead or too sick to attend. Only a year later Tyndall too was gone, due to a bizarre accident when his wife mistakenly administered an overdose of chloral to the ailing old man. Huxley had been frequently ill ever since 1885, and he found that attacking clerics was the most effective cure. He seemed to require controversy in order to convince not only his listeners of the value of scientific naturalism, but also himself.[4] In 1895 Huxley died in the middle of an attack on Balfour.

Spencer and Stephen lingered on into the twentieth century, outliving their friends, their vigor, and even their reputations.[5] A nervous breakdown in 1882 had made Spencer's task of completing the *Synthetic Philosophy* a struggle against physical and psychological deterioration. But the completion of his magnum opus in 1896 gave him no joy. Beatrice Webb speculated that Spencer "eventually discovered that there was no evidence in the findings of physical science for any such assumption of essential beneficence in the working of natural forces; and that the mental misery of his later life was not altogether unconnected with the loss of the inspiring creed with which he began his *Synthetic Philosophy.*"[6]

Spencer's nervous condition worsened as his philosophical system became subject to increasingly persistent and telling criticism.[7] Sensitive to shifts in the intellectual scene, he could detect the decline of his influence.[8] Indeed, after his death in 1903, his reputation had sunk so low that the Dean of Westminster could doubt that Spencer was sufficiently eminent to merit the national tribute of a memorial in Westminster Abbey.[9]

Sadder still was Spencer's attempt to find a place for immortality in his agnosticism as he wrestled with imminent death. In 1902 he published an essay on "Ultimate Questions" where he defended the scientific validity of immortality on the basis of the eternal and uncreated nature of space.[10] Morley rushed to Brighton, concerned that this represented a "weakening of Agnostic orthodoxy. It made some of the narrower or the firmer among us quake." Spencer had prepared some diagrams in order to demonstrate the geometric proofs for his position but Morley could only object that space is a subjective impression. Morley recollected that Spencer's eyes flashed as "he exclaimed, 'Then you have turned a Kantian, have you?' I saw that things could be carried no further, so with remorse in my heart I quitted him."[11]

The unexpected demise of Stephen's second wife in 1895 plunged him into the depths of a despair that seems to have lasted until his death. His daughter Virginia Woolf described Stephen as one "who, by the failure of some stay, reels staggering blindly about the world, and fills it with his woe." At meals Stephen sat oblivious to his family and "groaned aloud or protested again and again his wish to die."[12] Stephen began to feel isolated. To a friend he wrote in 1898 that "for reasons needless to mention, I am very lonely and often in very low spirits. . . . I am half afraid to say anything about myself—I don't want to appear morbid and yet I have little to say that can be called cheerful."[13] Stephen felt as though the world had passed him by.[14] If no letters arrived for him he would sigh "everyone has forgotten me."[15]

Like Spencer, Stephen also engaged in activities late in life which might have been thought previously to be at odds with his agnosticism. He was involved in the founding of the London Ethical Society in 1886.

He was president of the Ethical Society and also of the Rationalist Press Association, although Stephen had an aloof attitude and had no great faith in movements of any kind.[16] But Stephen contributed to the *Agnostic Annual* and delivered a number of addresses to ethical societies during the nineties. The addresses were published in *Social Rights and Duties* in 1896. Perhaps Stephen had seen that, like eighteenth-century deism, the agnostics had failed to capture the hearts and minds of the English, and now he became interested in seeing what could be accomplished through organized movements.

When Stephen died in 1904 agnosticism, as a distinct school of thought, was in its final death throes. The less important agnostics were either gone or interested in pursuing commitments other than agnosticism. Laing had died in 1897. Morley was busy with politics and his position as secretary of state for India (1905–10). Liberalism was his religion now. Gould, like Stephen, became attracted to organized freethought and later viewed agnosticism, as well as Positivism, socialism, and the Ethical Movement, as no longer adequate to meet the challenges of the twentieth century.[17] Even the secularists realized by 1907 that agnosticism was dead and that they no longer had anything to gain by using the term for their own ends. *The Agnostic Annual and Ethical Review* became *The Rationalist Press Association Annual and Ethical Review* in 1908.

Agnosticism did not long survive into the twentieth century as a specific movement in part because of vast upheavals in the intellectual framework of Europe. Symbiotically linked to the Victorian age and to the worship of science which swept all Europe in the latter half of the nineteenth century, agnosticism was eventually doomed to destruction when the forces of change which had created the society in which it flourished also transformed the world once more. The agnostic view of nature became obsolete when the new physics of Rutherford and Einstein revolutionized science in the second and third decades of the twentieth century. Even before the coming of Einstein, the so-called revolt against Positivism of the 1890s and the early years of the twentieth century was already under way. The mechanistic and naturalistic analysis of nature was subjected to attack by eminent thinkers from around the Western world, including the American William James, Henri Bergson and Henri Poincaré of France, the German Edmund Husserl, and James Ward of England.[18] By tying themselves to a particular view of nature so intimately connected to nineteenth-century society, ideas, and politics, the agnostics underestimated the power of science to evolve and grow. And since their god was deduced from their concept of nature it was only "natural" that the agnostic holy trinity became outdated and died a tragic death. With the advent of probabilistic laws in science, the dogma of uniformity in nature was no longer certain. A crisis of faith in the doctrine of cause was engendered by Heisenberg's uncertainty principle. It was not as easy to subscribe to the tenet of the existence of an external world now that the theory of relativity undermined traditional assumptions about the role of human beings as passive observers of an unchanging natural order.

Values which the agnostics had read into nature and which betray their Victorianism were also widely discredited. The belief in the rationality of people, the hopes for a world of peace, and the naive faith in the inevitability of progress, both material and moral, were all guaranteed by a view of nature which was no longer viable. The growth of rabid nationalism, social and economic dislocation, and the devastating impact of a world war spelled the end of the European cult of science.

But it was not just that the evolution of thought and reality passed the agnostics by—their creed was not easily disseminated to an English public. The *Christian Commonwealth*, in a review of the *Agnostic Annual*, pointed out in 1901 that agnosticism had "doomed itself to unpopularity" due to its learned but ponderous style of thought. "Agnosticism," the *Christian Commonwealth* announced, "as adumbrated in the earlier writings of MR. HERBERT SPENCER, promised to fascinate multitudes; but its later developments have been repellent to all but a tiny coterie, although many continue to read merely ethical magazines out of intellectual curiosity."[19] A philosophy bred from a cross of Kantian epistemology, German transcendentalism, and English empiricism could never be popular unless the simplifying genius of a host of Spencers continued to make it palatable to masses of converts. Stephen's fears that agnosticism would go the way of eighteenth century deism were well founded.

In addition to the unavoidable complexity of agnosticism there were difficulties with attempts at formulating a consistent theoretical framework. Dockrill argues that one of the reasons why agnosticism declined as a distinct school of thought was that the agnostics were not professional philosophers. "Generally," Dockrill claims, "agnosticism failed as a strictly philosophical movement."[20] The professional philosophers were opposed to them and contributed to the swelling body of criticism put forward during the nineties.

One area where the agnostics met with considerable opposition throughout their careers concerned their views on the relationship between science and religion. Critics, especially orthodox Christians, claimed that the agnostics purposely justified science at the expense of religion. We have seen that this approach to agnosticism does not do justice to the subtlety of their position. They genuinely believed that they preserved both science and religion in a peaceful harmony through the agnostic perspective, but the reconciliation was not a viable one despite the agnostics' best intentions. The upshot of their whole agnosticism was a pervasive scepticism. The validity of scientific first principles could not be justified scientifically or intellectually. The agnostics had limited knowledge to the extent where science could not ground itself knowingly, but only feelingly or faithfully. The validity of agnostic religion, in the intellectual realm, rested upon these same scientific axioms, so in the end religion, too, was forced to rely on feelings. Their agnosticism, at times, became an admission that neither science nor religion could be affirmed intellectually. To justify holding on to both they resorted to a position that retained science and religion, but admitted that our knowledge is too limited to allow us to perceive how to retain the two.

We have highlighted the genuine religious elements that do remain in the agnostics' thought through an examination of their debt to Mansel. Scientific naturalism and orthodox Christianity in the nineteenth century were like mirror opposites. An image in a mirror is always inverted and may at first glance be mistaken for a complete opposite, but the more basic similarity is revealed on closer examination. Scientific naturalism and orthodox Christianity shared a number of assumptions. An identical epistemological structure presented as many difficulties in reconciling science and religion for Mansel as it did for Huxley. There is a deep vein of scepticism in Mansel's thought, but at the same time he was dogmatic in his use of Kantian thought to defend the status quo and the infallibility of the biblical text as interpreted by the Anglican Church. The agnostics found it fruitful to borrow from Mansel, as they had a similar aim. Mansel's dogmatism is echoed in the agnostic attempt to dogmatically proclaim the infallibility of the text of nature as interpreted by the Church Scientific in order to protect conservative liberalism from the attacks of socialists and liberal democrats.[21]

Despite the end of agnosticism as a school of thought, the philosophical inconsistencies of Victorian agnosticism, and the reaction against Positivism during the late nineteenth and early twentieth centuries, the power of agnostic assumptions has continued up until the present day. Agnosticism proved to be extremely malleable. It could be adapted to the needs of varying intellectual movements. But what was retained by twentieth-century thinkers usually represented the shadow side of agnosticism. The optimistic, questioning, forward-looking, religious element in agnosticism was lost when the Western world entered what Baumer has called "The Age of Longing." The death of the agnostic god of science transformed the robust faith of agnosticism into a frustrated and despairing doubt longing for faith.

The twentieth century has spawned other varieties of agnosticism which also display a tragic quality. There are the complacent unbelievers who seize upon agnosticism in order to justify a self-serving, lazy attitude toward religious and metaphysical issues.[22] Maurice's fears that Mansel's Bampton Lectures would eventually lead to the stifling of religious questions proved to be valid.[23] There are also the aggressive unbelievers who cloak their atheism in a modified agnosticism and put forward a new gnosticism.[24]

But all agnosticism need not be tragic. Although the Huxleian variety of agnosticism is no more successful in attaining intellectual and moral integrity than the orthodoxy it attacked, including Mansel's version of agnosticism put forward as a defense of Christianity, this study has pointed to a third type of agnosticism represented by Kant which succeeds in justifying both science and religion intellectually and faithfully. Kant could demonstrate that the universal and necessary principles that ground science with certainty can be legitimized from the transcendental idealist position. The validity of Kant's approach depended on treating nature as appearance and limiting knowledge to the realm of appearances. But this same idea constituted the point of departure for Kant's ethical and religious thought. If knowledge is restricted to objects of possible experience, then people are not merely subject to the laws of nature, they are compelled by moral law to think of themselves as free beings, responsible for their actions and the values (gods) they choose to worship. A true defense of science must, paradoxically, also be a defense of religion. An authentic agnostic view of the limits of knowledge is a necessary component of a sound theism.

Kant's notion of two points of view could not be consistently adopted by Huxley and his colleagues. The agnostic stress on causal determinism endangered the notion of freedom upon which Kant had built his justification of religion and moral "proof" of God.[25] Kant's work shows us the possibility and importance of developing a perspective that allows us to appropriate knowledge (of nature) in order to show that the house we dwell in is that of faith (Kant's practical reason).

Our historical study of agnosticism has, if nothing else, revealed the tremendous flexibility in the agnostic principle. Perhaps we can mold agnosticism in such a way that it is no longer destructive, pessimistic, or tragic. The potential exists for a reinterpretation of our understanding of agnosticism which would allow for the return of religious and spiritual elements. Agnosticism of the Kantian variety curtails dogmatism, defeats scepticism, and can contribute something of value to all systems of belief. The challenge of the future is to see if a healthy agnosticism, which actively questions everything, including itself, can be rescued from the remains of tragic agnosticism.

ABBREVIATIONS

The following abbreviations are used frequently in the notes and, at times, in references in the text. They are listed here in alphabetical order by abbreviation:

AA	Leslie Stephen, *An Agnostic's Apology and Other Essays* (London: Smith, Elder & Co., 1893; reprint ed., Westmead, Farnborough, Hants: Gregg International Publishers, 1969)
CPR	Immanuel Kant, *Critique of Pure Reason,* trans. Norman Kemp Smith (New York: St. Martin's Press, 1965)
DPL	Sir William Hamilton, *Discussions on Philosophy and Literature, Education and University Reform* (London: Longman, Brown, Green & Longmans; Edinburgh: Maclachlan & Stewart, 1853)
EE	T. H. Huxley, *Evolution and Ethics and Other Essays* (New York: D. Appleton & Co., 1899)
FP	Leslie Stephen, *Essays on Freethinking and Plainspeaking* (London: Longmans, Green & Co., 1873)
FPNSP	Herbert Spencer, *First Principles of a New System of Philosophy* (New York: D. Appleton & Co., 1882)
FS	John Tyndall, *Fragments of Science* (London: Longmans, Green & Co., 1892)
HETEC	Leslie Stephen, *History of English Thought in the Eighteenth Century* (London: Smith, Elder & Co., 1876)
ICST-HP	London, Imperial College of Science and Technology, The Huxley Papers
LE	William Kingdon Clifford, *Lectures and Essays,* ed. Leslie Stephen and Frederick Pollock (London: Macmillan & Co., 1879)
LLHS	David Duncan, ed., *Life and Letters of Herbert Spencer* (London: Methuen & Co., 1908)
LLLS	Frederic William Maitland, *The Life and Letters of Leslie Stephen* (London: Duckworth & Co., 1906)
LLTHH	Leonard Huxley, ed., *Life and Letters of Thomas Henry Huxley* (New York: D. Appleton & Co., 1900)

LML Sir William Hamilton, *Lectures on Metaphysics and Logic*, ed. Rev. H. L. Mansel and John Veitch (Stuttgart-Bad Cannstatt: Friedrich Frommann Verlag, 1970)

LPK Henry Longueville Mansel, *A Lecture on the Philosophy of Kant* (Oxford: John Henry & James Parker, 1856)

LRT Henry Longueville Mansel, *The Limits of Religious Thought Examined in Eight Lectures*, 4th ed. (London: John Murray, 1859)

LWJT Arthur Stewart Eve and C. H. Creasey, *Life and Work of John Tyndall* (London: Macmillan & Co., 1945)

MR T. H. Huxley, *Method and Results* (New York: D. Appleton & Co., 1897)

NF John Tyndall, *New Fragments* (New York: D. Appleton & Co., 1896)

PC Henry Longueville Mansel, *The Philosophy of the Conditioned* (London and New York: Alexander Strahan, 1866)

PL Henry Longueville Mansel, *Prolegomena Logica* (Oxford: William Graham, 1851)

PP Herbert Spencer, *The Principles of Psychology* (London: Longman, Brown, Green & Longmans, 1855)

RI-TP London, The Royal Institution, The Tyndall Papers

SCT T. H. Huxley, *Science and Christian Tradition* (London: Macmillan & Co., 1909)

SE T. H. Huxley, *Science and Education* (New York & London: D. Appleton & Co., 1914)

SHT T. H. Huxley, *Science and Hebrew Tradition* (New York: D. Appleton & Co., 1898)

NOTES

Introduction

1. Franklin L. Baumer, *Religion and the Rise of Scepticism*, 11.
2. Julian Huxley et al., *Science and Religion*, 78.
3. Ibid.
4. Richard H. Popkin, *High Road to Pyrrhonism*, 71–72.
5. I must make it clear here that I do not see myself as a neo-Kantian, slavishly dependent on my master Kant. I am not, like the late-nineteenth-century German intellectuals, proclaiming that we should go "back to Kant." Neither would I endorse any call to find truth in one single author, whether it be "back to Marx" or "back to Weber." I would see in my interpretation of Kant the beneficial traces of my reading of Hegel, Kierkegaard, Marx, Nietzsche, Dilthey, Weber, Buber, Heisenberg, Eddington, and Barfield, among others.
6. Dominick LaCapra, "Rethinking Intellectual History and Reading Texts," 47–85.

Chapter One
The Agnostic Conundrum

Epigraph from Stephen, *AA*, 8–9.

1. Two editions appeared almost immediately in 1858, two more in 1859, and a fifth was published in 1867.
2. Anon., "Mansel's Bampton Lectures," 10.
3. Mrs. Charles Lyell, ed., *Life, Letters and Journals of Sir Charles Lyell* 2:321–22. See also Leonard Huxley, ed., *LLTHH* 1:234–35.
4. Ronald Paulson, *Hogarth: His Life, Art, and Times*, 2:224.
5. T. H. Huxley, "Mr. Balfour's Attack on Agnosticism," 534.
6. Andrew Seth, *Scottish Philosophy*, 180.
7. E. Marshall, "Agnosticism," 489; V.H.I.L.I.C.I.V., "Agnosticism," 34; E. H. M. and J. R. Thorne, "Agnosticism: Agnostic," 418.
8. James A. H. Murray, ed., "Agnostic," 186. It is generally accepted that Huxley coined the word *agnosticism*. A description of a brief but inconclusive controversy questioning this point can be found in the *Times Literary Supple-*

ment: John W. Bicknell, ''Neologizing'' (29 June 1973), 749; Robert H. Tener, ''Neologizing'' (10 Aug. 1973), 931; Byron Farwell, ''Neologizing'' (31 Aug. 1973), 1002–3; Robert H. Tener, ''Neologizing'' (9 Nov. 1973), 1373.

9. Robert H. Tener, ''R. H. Hutton and 'Agnostic,' '' 429–31; Robert H. Tener, ''Agnostic,'' 732. The earliest appearance of the word *agnostic* in print is Hutton's ''Theological Statute at Oxford,'' 642.

10. Murray, ed., ''Agnostic,'' 186.

11. The present O.E.D. still asserts that Huxley took the term *agnostic* from Acts. The following scholars repeat Hutton's tale: Jacob Gould Schurman, *Agnosticism and Religion*, 85; Robert Flint, *Agnosticism*, 2; Alfred William Benn, *History of English Rationalism in the Nineteenth Century*, 2:199; George W. Hallam, ''Source of the Word 'Agnostic,' '' 265–69; L. E. Elliot-Binns, *English Thought, 1860–1900*, 324.

12. It was not until 1947 that the authenticity of Hutton's account was questioned by Onions, who recognized the significance of the etymological discrepancy. See C. T. Onions, ''Agnostic,'' 225.

13. We can also now deal with two important items having reached this point in the discussion. First, some scholars have claimed that the term *agnostic* is linguistically incorrect; however, they reach this conclusion since they assume that Hutton's story is authentic. (For example, see Alfred E. Garvie, ''Agnosticism,'' 1:215.) To connect *agnostic* with the Greek word used in Acts 17:23 to mean ''to an Unknown God'' does not make etymological sense. But this whole difficulty is avoided if one sees *agnostic* as derived from *gnostic*. See Onions, ''Agnostic,'' 225. Second, we can now fix the date of origin more accurately. Murray's claim (in *Notes and Queries*, 6th ser., 6, [18 Nov. 1882], 418) that Huxley invented the term in September of 1869 is incorrect because Hutton used the word *agnostic* in an article published 29 May 1869. Huxley seems to indicate that he coined the term when the Metaphysical Society had already begun its meetings, and Hutton claims that the event took place before the formation of the society, but there is no real conflict here between the two stories. On 21 April 1869, there was an organizational meeting of the society attended by both Hutton and Huxley. The next meeting, the first during which a paper was read (Hutton on ''Mr. Herbert Spencer's Theory of the Gradual Transformation of Utilitarian into Intuitive Morality by Hereditary Descent'') took place on 2 June 1869. It seems safe to say that Huxley coined the term shortly after the organizational meeting held on April 21st at Willis's Rooms, that he considered that meeting to be the first session, and that Hutton rather promptly put the word into print about a month later but saw the proceedings of June 2d as the real starting point of the society. (If we are to accept Hutton's claim that Huxley presented the new term at Knowles's house, then the April 21st meeting at Willis's Rooms would be ruled out as the date of origin.) Why Hutton maintained that *agnostic* originated from the inscription on the Athenian altar of Acts is not clear, but perhaps he read this into Huxley's words on the basis of a reference to the altar of the ''Unknown and Unknowable'' in one of Huxley's essays published earlier in the decade. (''On the Advisableness of Improving Natural Knowledge,'' *Fortnightly Review* 3 [Jan. 1866], 636.)

14. Leslie Stephen also saw the agnostic-gnostic opposition as important for explaining the meaning of agnosticism. See Stephen, *AA*, 2.

15. Tyndall hinted at the inconsistency of modern Christianity in a similar fashion. "Then arose the sect of Gnostics,—men who *know*—who laid claim to the possession of a perfect science, and who, if they were to be believed, had discovered the true formula for what philosophers called the Absolute. But these speculative Gnostics were rejected by the conservative and orthodox Christians of their day as fiercely as are their successors the Agnostics,—men who *don't know*,—are rejected by the orthodox in our own." Tyndall, *NF*, 9.

16. James Martineau, *Study of Religion* 1:xi.

17. Rev. Thomas Corbishley, S. J., "Agnosticism," 49 (Heraclitus and Protagoras); Flint, *Agnosticism*, 86 (Protagoras and Gorgias); Walter Kaufmann, *From Shakespeare to Existentialism*, 69 (Socrates); Flint, *Agnosticism*, 42, 95 (Carneades and Sextus Empiricus); R. J. Zwi Werblowsky and Geoffrey Wigoder, "Agnosticism," 16 (Maimonides); T. Gilby, U. Voll, P. K. Meagher, "Agnosticism," 77 (Maimonides); Flint, *Agnosticism*, 98 (Occam and Peter D'Ailly), 526 (Luther), 102 (Agrippa), 527 (Faustus Socinus), 104 (Montaigne), 107 (Charron), 114 (Pascal), 113 (Huet), 116 (Bayle), 531 (King and Browne); J. O'Higgins, "Browne and King, Collins and Berkeley," 90 (King and Browne); Flint, *Agnosticism*, 527 (Hutchinson); Baumer, *Religion and the Rise of Scepticism*, 144 (Hume); James Noxon, "Hume's Agnosticism," 361–83 (Hume); John Passmore, "Darwin's Impact on British Metaphysics," 45 (Hume); Henry Calderwood, "Agnosticism," 37 (Kant); Kaufmann, *From Shakespeare to Existentialism*, 69 (Goethe); Flint, *Agnosticism*, 532 (Schleiermacher); Frederick Copleston, *History of Philosophy*, Vol. 8, pt. 1, 302 (James Mill); John Theodore Merz, *History of European Thought in the Nineteenth Century* 3:315 (James Mill); Flint, *Agnosticism*, 42 (Lamennais); W. R. Sorley, *History of British Philosophy to 1900*, 248 (Hamilton); Calderwood, "Agnosticism," 37 (Hamilton); Frank M. Turner, "Victorian Scientific Naturalism and Thomas Carlyle," 336 (Carlyle); H. J. Blackham, "Introduction—Humanism: The Subject of the Objections," in *Objections to Humanism*, 12 (Comte); Baumer, *Religion and the Rise of Scepticism*, 144 (Comte); Garvie, "Agnosticism," 216 (Comte); Clement C. J. Webb, *Study of Religious Thought in England from 1850*, 83 (J. S. Mill); A. O. J. Cockshut, *Unbelievers*, 19–30 (J. S. Mill); J. M. Robertson, *History of Freethought in the Nineteenth Century* 1:296 (Holyoake); Cockshut, *Unbelievers*, 31–43 (Clough), 44–58 (Eliot); Passmore, "Darwin's Impact on British Metaphysics," 45 (Mansel); Sorley, *History of British Philosophy to 1900*, 248 (Mansel); Flint, *Agnosticism*, 561 (Mansel); Calderwood, "Agnosticism," 37 (Mansel); Richard Holt Hutton, "Moral Significance of Atheism," 22 (Tyndall); [William Barry], "Professor Huxley's Creed," 160 (Tyndall); James G. Paradis, *T. H. Huxley*, 104 (Tyndall); Clarence Ayres, *Huxley*, 110 (Tyndall); Eve & Creasey, *LWJT*, 283 (Tyndall); Charles S. Blinderman, "John Tyndall and the Victorian New Philosophy," 286 (Tyndall); Joe D. Burchfield, "John Tyndall—A Biographical Sketch," in Brock et al., eds., *John Tyndall*, 7 (Tyndall); Ronald W. Clark, "Tyndall as Mountaineer," in Brock et al., eds., *John*

Tyndall, 67 (Tyndall); D. W. Dockrill, "Origin and Development of Nineteenth Century English Agnosticism," 4 (Tyndall); Cockshut, *Unbelievers*, 73–85 (Spencer); Dennis R. Dean, " 'Through Science to Despair': Geology and The Victorians," in Paradis and Postlewait, eds., *Victorian Science and Victorian Values*, 126 (Arnold); Richard A. Armstrong, *Agnosticism and Theism in the Nineteenth Century*, 96 (Arnold); J. B. Bury, *History of Freedom of Thought*, 218 (Arnold); Flint, *Agnosticism*, 533 (Ritschl); W. T. Davison, "Poetic Agnosticism," 128 (Meredith); Bury, *History of Freedom of Thought*, 214 (Stephen); Cockshut, *Unbelievers*, 99–114 (Butler); Phyllis Grosskurth, *Leslie Stephen*, 22 (Swinburne); Davison, "Poetic Agnosticism," 128 (Swinburne); Owen Chadwick, *Victorian Church*, pt. 2, 140 (Sidgwick); Flint, *Agnosticism*, 536 (Sabatier); Robertson, *History of Freethought in the Nineteenth Century*, 549 (James); Leon Stephen Jacyna, "Scientific Naturalism in Victorian Britain," 89 (Clifford); Paradis, *T. H. Huxley*, 104 (Clifford); Rev. A. W. Momerie, *Agnosticism*, 124 (Clifford); Alexander Macfarlane, "William Kingdon Clifford," 91 (Clifford); Flint, *Agnosticism*, 426 (Clifford); Dockrill, "Origin and Development of Nineteenth Century English Agnosticism," 4 (Clifford); Robertson, *History of Freethought in the Nineteenth Century* 2:547 (Bradley); Cockshut, *Unbelievers*, 146 (Ward); Frederic R. Crownfield, "Whitehead: From Agnostic to Rationalist," 377 (Whitehead); M. R. Holloway, "Agnosticism," 1:205 (Russell); Blackham, "Introduction—Humanism," in *Objections to Humanism*, 14 (Buber, Jaspers, and Marcel).

18. T. H. Huxley, *Hume* (1879), 58; Huxley, *SCT*, 237.

19. Although short pieces in encyclopedias and dictionaries offer analyses of agnostic epistemology, larger works are strangely lacking. Cockshut's *Unbelievers* does not deal with the epistemological foundations of agnosticism. In his *Agnosticism*, Robert Flint emphasizes that agnosticism is primarily epistemological, but his rather loose definition of the doctrine leads him to find it throughout history, thus denying the distinctive flavor of Victorian agnosticism. D. W. Dockrill, in his article "The Origin and Development of Nineteenth Century English Agnosticism," succeeds in examining the epistemological dimension of Victorian agnosticism in the context of an analysis of how this new form of scepticism was grounded in the Victorian ethos.

20. Tyndall and Clifford are not always considered by scholars to be agnostics. Tyndall is often referred to as a materialist (see Bernard M. G. Reardon, *From Coleridge to Gore*, 297) and Clifford, according to Cockshut, was a dogmatic atheist who represented no one but himself (Cockshut, *Unbelievers*, 67, 123). With the exception of one article and an unpublished dissertation, there exist no detailed studies of Huxley, Stephen, Clifford, Tyndall, and Spencer as a group. See D. W. Dockrill, "Origin and Development of Nineteenth Century English Agnosticism," 3–31; Dockrill, "Studies in Nineteenth-Century English Agnosticism."

21. Ulke is so impressed by Hamilton's "philosophy of the conditioned" that in his recently published book he portrays the Scottish philosopher as the prime agnostic. See Karl-Dieter Ulke, *Agnostisches Denken im Viktorianischen England*. Strangely enough, Ulke sees agnosticism ending in 1865 when Mill's *Examination of Sir William Hamilton's Philosophy* supposedly de-

stroyed Hamilton's credibility. This would place the end of agnosticism four years before Huxley coined the term. Scholars who have briefly noted the Kant-Mansel-agnosticism link include Garvie, "Agnosticism," 214–20; Holloway, "Agnosticism," 1:205–9; Ronald W. Hepburn, "Agnosticism," 1:56–59; Corbishley, "Agnosticism," 49–52; Calderwood, "Agnosticism," 36–39; M. M. Waddington, *Development of British Thought from 1820 to 1890*, 129; Claude Welch, *Protestant Thought in the Nineteenth Century*, 254; Gilby et al., "Agnosticism," 77–78; Sorley, *History of British Philosophy to 1900*, 243, 248; Benn, *History of English Rationalism* 1:37. Since many of these pieces are short encyclopedia articles they are understandably superficial. More substantial treatments are to be found in Ward's *Naturalism and Agnosticism*, and Flint's *Agnosticism*, but both are somewhat outdated. Dockrill's "Origin and Development of Nineteenth Century English Agnosticism" is strong on the Mansel-agnosticism link but says little about Kant. Cockshut does not mention Kant or Mansel once throughout *The Unbelievers*.

22. Benn, *History of English Rationalism* 2:453. Among those who do not see Christian agnosticism as a contradiction in terms are Gilby et al., "Agnosticism," 77; Schurman, *Agnosticism and Religion*, 64; Cockshut, *Unbelievers*, 92–93.

23. Henry Wace, *Christianity and Agnosticism*, 2, 6. See also Anon., "Prevalent Phase of Unbelief," 167; Anon., "Popular View of Atheism," 819.

24. Frederick Engels, *Socialism Utopian and Scientific*, 13; V. I. Lenin, *Materialism and Empirio-criticism*, 195. A number of twentieth-century scholars have agreed with Wace and Engels that agnosticism was indistinguishable from atheism. See Eileen Barker, "Thus Spake the Scientist," 90; Blackham, "Introduction—Humanism," in *Objections to Humanism*, 14.

25. RI-TP, British Correspondence of John Tyndall at the Royal Institution, 3413. All letters, notebooks, and manuscripts from this collection will also be accompanied by the citation recommended by Friday et al. in *John Tyndall, Natural Philosopher*—R. I. MSS T., 20/C7, 52.

26. Huxley, *EE*, 134; Leonard Huxley, ed., *LLTHH* 1:259–60; ibid. 2:172; T. H. Huxley, "Mr. Huxley's Doctrine," 158; Stephen, *AA*, 1.

27. Webb, *Study of Religious Thought*, 83; Charles Singer, *Religion and Science Considered in Their Historical Relations*, 77; Copleston, *History of Philosophy*, 302; Kai Nielsen, "Agnosticism," 17; Cockshut adheres to this view throughout *The Unbelievers*.

28. I shall adopt Baumer's definition of religion here as a belief system that includes a transcendental element or a "concern for the *metaphysical* overtones of human life." (Baumer, *Religion and the Rise of Scepticism*, 29.) This approach avoids a narrow definition which would restrict religion to a particular set of dogmas, and it avoids a too-broad definition of religion as devotion to any end outside the individual.

29. Momerie, *Agnosticism*, 3. See also Webb, *Study of Religious Thought*, 70; Franklin Baumer, *Modern European Thought*, 355; R. L. Franklin, "Religion and Religions," 420; Benn, *History of English Rationalism* 2:386–87; Corbishley, "Agnosticism," 51; Gerald Birney Smith, "Agnosticism," 9.

30. Hutton, "Moral Significance of Atheism," 23. On the possibility of re-

ligious agnosticism see also Ronald W. Hepburn, "A Critique of Humanist Theology," *Objections to Humanism*, ed. H. J. Blackham, 50; Ronald Hepburn, "Gospel and the Claims of Logic," in *Religion and Humanism*, Ronald Hepburn et al. 18; Hepburn, "Agnosticism," 57; Flint, *Agnosticism*, 401.

31. Including the agnostics within the sceptical tradition raises a number of tricky questions. The term *scepticism* itself is rather vague, grouping together those who doubt or disbelieve generally accepted ideas. To be more specific and label the agnostics as religious sceptics, as Baumer has in *Religion and the Rise of Scepticism*, is awkward because it could apply equally to the fideists, who were sceptics but used sceptical arguments to support Christianity. However, to reserve irreligious sceptics for agnosticism is misleading, because this implies that the agnostics were antireligious. Philosophical scepticism, although capturing the importance of epistemology to the agnostics, fails to indicate their application of the notion of the limits of knowledge to a concept of God and the extravagances of Victorian theologians. The complexity of agnosticism eludes categorization here because the agnostics cannot be termed religious sceptics, irreligious sceptics or philosophical sceptics. I have therefore chosen to speak of agnosticism as a unique form of scepticism although it still strikes me as problematic.

32. Flint, *Agnosticism*, 4.

33. Other contemporary works of the same first-rate quality are James Ward's *Naturalism and Agnosticism* (1899) and Jacob Gould Schurman's *Agnosticism and Religion* (1896). Not as brilliant but still useful is A. W. Momerie's *Agnosticism* (1884). Richard A. Armstrong's *Agnosticism and Theism in the Nineteenth Century* (1905) and Henry C. Sheldon's *Unbelief in the Nineteenth Century* (1907) are not, despite their titles, major works on agnosticism.

34. Richard H. Popkin, *History of Scepticism*, ix.

35. John W. Bicknell, "Leslie Stephen's 'English Thought in the Eighteenth Century,'" 107–8; Owen Chadwick, *Secularization of the European Mind in the Nineteenth Century*, 153.

36. Carl L. Becker, *Heavenly City of the Eighteenth-Century Philosophers*, 29.

37. Huxley, *SCT*, 41; T. H. Huxley, "Science and Religion," 36.

38. Stephen voiced similar reservations about Enlightenment deism. He referred to constructive deism as the attempt to "substitute for Christianity a pure body of abstract truths, reposing on metaphysical demonstration." Deism decayed due to its internal weakness. "The metaphysical deity was too cold and abstract a conception to excite much zeal in his worshippers." Stephen, *HETEC* 1:169. Bicknell has pointed out that Stephen's ambivalence toward the philosophes in his *History of English Thought* stems from his attempt to simultaneously demonstrate that orthodoxy had been destroyed in the last century and learn why deism had failed to capture the hearts and minds of the English. Bicknell, "Leslie Stephen's 'English Thought in the Eighteenth Century,'" 108, 112, 118–19. See also Floyd Clyde Tolleson, Jr., *Relation Between Leslie Stephen's Agnosticism and Voltaire's Deism*. Tolleson argues that "Voltaire, among other influences, worked to make Leslie

Stephen an agnostic" (2). Although Tolleson establishes the importance of Voltaire's thought for Stephen, he very often ignores the "other influences," including Mansel and the Kantian tradition.

39. This will no doubt strike some readers as a rather bold assertion in light of the fact that Huxley often referred to Hume as an agnostic forefather and contributed a piece of hagiography to Morley's English Men of Letters series entitled *Hume*, not *Kant*. In addition to this, Hume emerges as the hero of Stephen's *English Thought* (See Bicknell, "Leslie Stephen's 'English Thought in the Eighteenth Century,'" 120.) But a careful reading of Huxley's *Hume* will reveal not only Huxley's high regard for Kant but his tendency to supplement Hume's shortcomings in epistemology with Kantian notions. See Huxley, *Hume*, 65, 85. In a discussion on the problem of innate ideas, Huxley resolves the issue by bringing in Kant's "doctrine of the existence of elements of consciousness, which are neither sense-experiences nor any modifications of them." (Huxley, *Hume*, 85.) Huxley's interest in Kant's notion of the structure of the mind, an approach that Hume did not take despite his emphasis on the limits of knowledge, explains Huxley's desire to use Kant as a fruitful response to difficulties in Humean epistemology. It is striking to find a similar handling of Hume by Stephen. After discussing the impossibility of building a consistent philosophy of natural science on Hume's principles, Stephen turns to Kant (and Spencer) as necessary avenues of escape from the destructiveness of Hume's scepticism. (Stephen, *HETEC* 1:48–56.)

40. Dockrill, "Origin and Development of Nineteenth Century English Agnosticism," 3–31; Dockrill, "T. H. Huxley and the Meaning of 'Agnosticism,'" 461–77; Dockrill, "Studies in Nineteenth Century English Agnosticism."

41. T. H. Huxley, *Lay Sermons, Addresses, and Reviews*, 150. As late as 1903 Flint had to remind his contemporaries that "a very common misconception as to agnosticism is that it is identical with positivism." (Flint, *Agnosticism*, 52.)

42. W. M. Simon, *European Positivism in the Nineteenth Century*, 4.

43. Huxley, *MR*, 156, 158. Huxley attacked Positivism many times after his articles of the late 1860s. See Huxley, *SCT*, 211; Leonard Huxley, ed., *LLTHH* 2:244. Huxley was particularly outraged by the authoritarian strain in Positivist thought, which not only led to Comte's belief that science should be regulated by the state, but also seemed to Huxley to spill over into Comte's idea for a new religion. Huxley branded the religion of humanity as "spiritual tyranny and slavish social 'organization.'" (T. H. Huxley, "An Apologetic Irenicon," 559.) For more on Huxley's rejection of Comte see Sydney Eisen, "Huxley and the Positivists," 337–58.

44. ICST-HP 23:12. (Morley to Huxley, 13 Jan. 1869.)

45. ICST-HP 8:69.

46. For a defense of a strong epistemological connection between Comte and the agnostics see Baumer, *Religion and the Rise of Scepticism*, 145–46.

47. Huxley, *Hume*, 50.

48. Huxley, *MR*, 144. See also Frank M. Turner, "Lucretius among the Victorians," 329–48.

49. ICST-HP 8:80; Huxley, *MR*, 155.

50. Tyndall, *FS* 2:191; Eve and Creasey, *LWJT*, 160; Stephen, *AA*, 128; Leonard Huxley, ed., *LLTHH* 2:154.

51. Many scholars still consider Huxley, Tyndall, and Clifford to be materialists. See Reardon, *From Coleridge to Gore*, 297; Maurice Mandelbaum, *History, Man, and Reason*, 23; Henry C. Sheldon, *Unbelief in the Nineteenth Century*, 60. For a rejection of the view of Huxley and Tyndall as materialists see Charles S. Blinderman, "T. H. Huxley," 50–62; idem, "John Tyndall and the Victorian New Philosophy," 288.

52. Dr. Louis Büchner, *Force and Matter*, 2, 29; Huxley, *EE*, 131; Frederick Gregory, *Scientific Materialism in Nineteenth-Century Germany*, 146–48; Huxley, *MR*, 162; Huxley, *EE*, 132.

53. Huxley, *MR*, 164; Leonard Huxley, ed., *LLTHH* 1:262; Tyndall, *FS* 2:72–73; Leslie Stephen, *Social Rights and Duties* 2:211; Stephen, *FP*, 89; Clifford, *LE* 2:58; Clifford, *Seeing and Thinking*, 90.

54. Gregory distinguishes between the German scientific materialists and what he calls a group of reductionists (such as Helmholtz, Ludwig, and Du Bois-Reymond). The latter group resembled the English agnostics in retaining materialism only as a maxim of scientific research. (See Gregory, *Scientific Materialism in Nineteenth Century Germany*, 149.) For example, Emil Du Bois-Reymond, the Berlin physiologist, held that the concepts of matter and force were only abstractions from natural phenomena which yielded no final explanation. Some problems, Du Bois-Reymond argues in his "The Limitation of Natural Knowledge" (1872), are forever beyond human knowledge.

55. Engels, *Socialism Utopian and Scientific*, 15; Lenin, *Materialism and Empirio-criticism*, 114.

56. Stephen, *HETEC* 1:34. Stephen repeated his point in a later work. "Englishmen were practically, if not avowedly, predisposed to empiricism." Leslie Stephen, *English Utilitarians* 3:77. Other scholars have said that empiricism is England's national school by virtue of its status as the prevailing outlook of English society, especially in the nineteenth century. See Susan Budd, *Varieties of Unbelief*, 270; John Herman Randall, Jr., *Career of Philosophy* 1:583; Dr. Rudolf Metz, *A Hundred Years of British Philosophy*, 47; Benn, *History of English Rationalism* 1:203. The conditions of life in the Victorian period, Houghton has asserted, could only have heightened the characteristic empiricist quality of English thought. See Walter Houghton, *Victorian Frame of Mind, 1830–1870*, 110–11.

57. Stephen, *English Utilitarians* 3:76; Michael Ruse, *Darwinian Revolution*, 145; H. L. Stewart, "J. S. Mill's 'Logic,'" 369.

58. Engels, *Socialism Utopian and Scientific*, 13.

59. Vernon F. Storr, *Development of English Theology in the Nineteenth Century, 1800–1860*, 326.

60. Stephen, *The Science of Ethics*, 358.

61. Stephen, *HETEC* 1:311. Years later, in 1903, Stephen was less harsh on Mill and described his religious influence as "latent," for although Mill had

never publicly stated his beliefs, his philosophy implicitly led to agnosticism. See Leslie Stephen, *Some Early Impressions*, 76.

62. John Stuart Mill, *Autobiography*, 44.

63. A.O.J. Cockshut's book *The Unbelievers* really does not distinguish between the unbelief of the men of the forties and fifties and the agnosticism of a later period. Defining agnostics loosely as those who rejected orthodox Christianity, but who were not totally antireligious like the atheists and secularists, allows Cockshut to include studies of Clough, J. S. Mill, Matthew Arnold, George Eliot, Samuel Butler, Huxley, and Spencer. Froude is dealt with as an agnostic, and Clifford is treated as a dogmatic atheist (p. 67). Cockshut chooses to dwell on how the thinkers' personal tempers and characters influenced their religious thought. There is virtually no discussion of the agnostic notion of the limits of knowledge. *The Unbelievers* is a disappointing work, especially since it is one of the few major secondary sources specifically on agnosticism.

64. Arthur Hugh Clough, *Poems and Prose Remains of Arthur Hugh Clough* 1:295.

65. Frank Turner, *Between Science and Religion*, chap. 2 on "Victorian Scientific Naturalism"; idem, "Rainfall, Plagues, and the Prince of Wales," 46–65; idem, "Victorian Conflict between Science and Religion," 356–76; idem, "Public Science in Britain, 1880–1919," 589–608; idem, "Victorian Scientific Naturalism and Thomas Carlyle," 325–43; idem, "John Tyndall and Victorian Scientific Naturalism," in Brock et al., eds., *John Tyndall*, 169–80. To this list of Turner's works can be added Leon Stephen Jacyna's "Scientific Naturalism in Victorian Britain," 1980, a fine thesis that builds on Turer's approach.

66. A list of scientific naturalists includes Positivists (e.g., Harrison) and other unbelievers who were not specifically agnostics (Lewes, Tylor, Lubbock, Lankester, Maudsley, Allen). See Hock Guan Tjoa, *George Henry Lewes*, 102; [Leslie Stephen], "George Henry Lewes (1817–1878)," *The Dictionary of National Biography*, ed. Sir Leslie Stephen and Sir Sidney Lee (Oxford: Oxford University Press, 1917), 11:1044–45 (Stephen labels Lewes a Positivist); Edward Clodd, *Grant Allen: A Memoir*, 192 (Clodd quotes Allen's denial that he is an agnostic).

67. Robert M. Young, "Historiographic and Ideological Contexts of the Nineteenth-Century Debate on Man's Place in Nature," in Young, *Darwin's Metaphor*, 240. Leon Jacyna's dissertation, already cited, and the work of Frank Turner, also follow up the social history of ideas approach. Older studies have noticed the affiliation of agnosticism with the Victorian middle class. Benn asserted that "agnosticism, its meaning once grasped or even dimly suspected, seemed well suited to the generally businesslike and sensible character of the English middle-class." (Benn, *History of English Rationalism* 1:203.) Noel Annan has included agnostics such as Darwin, Huxley, and Stephen in his study of the rise of a middle-class intellectual aristocracy in mid-nineteenth-century England. See Noel Gilroy Annan, "Intellectual Aristocracy," 243–87. But the recent work by Turner, Young, and Jacyna is more sophisticated in its treatment of the social context of unbelief.

68. I am indebted to James Moore for this insight.

69. In a comparison of scientific naturalism in mid-Victorian England and French scepticism in the late eighteenth century, Jacyna finds the novelties of the former to be tied to the special features of nineteenth-century British society. (Leon Jacyna, "Scientific Naturalism in Victorian Britain," 11.) However, I would argue that there were unique intellectual as well as social factors in scientific naturalism. It is important to avoid two dangers here. First, we must not reduce the intellectual content of agnosticism, as if we were vulgar Marxists, to the social context. Although the agnostics were molded by the socio-economic and political structure, they were individuals who acted upon and transformed that structure, in part through the way they shaped the future course of unbelief. Second, the notion of intellectual influence presents similar difficulties. By pointing to the impact of Mansel on agnosticism I do not mean to put forward a crude notion of influence wherein the agnostics have no active role. They were intelligent, independent men who, although impressed by the Kantian tradition, reformulated it into a distinctive new philosophical position.

70. Recently, James Turner has argued that American agnosticism arose due to the increasing willingness of liberal Christian leaders to adapt their religious beliefs to modernity. American Christianity became so rationalized that it decayed from within. However, my approach is quite different in that I trace the origins of English agnosticism to an affinity with orthodox Christian thinkers like Mansel who stressed the transcendence of God. See James Turner, *Without God, Without Creed.*

71. A number of authors have pointed to science in general (and not to Darwin specifically), as a crucial factor in the birth of agnosticism. Huxley and agnosticism are discussed under the chapter heading "The Scientific Movement," in Waddington, *Development of British Thought*, 124. In a bibliographical essay entitled, "The Unbelievers," Bicknell asserts that Benn's *History of English Rationalism* and Robertson's *History of Freethought in the Nineteenth Century* are among the many books that emphasize the notion that unbelief arose from developments in science. See Bicknell, "Unbelievers," 474, 484. Bury discusses the triumph of rationalism in the context of the advance of science. See Bury, *History of Freedom of Thought*, 226. As Wolff remarks, "it has become a truism that science—along with 'higher criticism' of the Bible—made doubters out of believers," but he later rejects the statement. See Robert Lee Wolff, *Gains and Losses*, 419. See also Baumer, *Religion and the Rise of Scepticism*, 93, 144.

The problematic nature of Darwinian theory for Christian theology has led scholars to treat evolution as one of the chief factors in the unsettlement of faith. Passmore affirms that there was a "natural alliance" between "Darwin and agnosticism" for "by destroying the argument from design Darwin did not disprove God's existence, but cut away the only argument in its favour which had any appeal to those who accepted the positivist doctrine that all knowledge derives from the observation of natural processes." (Passmore, "Darwin's Impact on British Metaphysics," 46.) Historical studies discuss the birth of agnosticism under headings and within a context which can only imply that a strong connection exists between evolutionary theory and all who subscribed to Huxley's position; the studies are in chapters with titles such as "The Evolutionary-Naturalist School," "Reactions to Darwin," "Evolution and Philosophy," "The Theory of Evolution," "Herbert Spencer and the Philosophy of Evolution," and "The Theory of Evolution." See Metz, *A Hundred Years of British Philosophy*, 111; Roland N. Stromberg, *Intellectual History of Modern Europe*, 312; Sir William Cecil Dampier, *A History of Science and Its Relations with Philosophy and Religion*, 318; D. C. Somervell, *English Thought in the Nineteenth Century*, 132; Sorley, *History of British Philosophy to 1900*, 274; Reardon, *From Coleridge to Gore*, 296. See also Richard D. Altick, *Victorian People and Ideas*, 230; Basil Willey, *Christianity Past and Present*, 109.

Alongside the destructive impact of science and evolutionary theory, scholars have placed the findings of biblical criticism as an important factor in the growth of nineteenth-century English unbelief. See Wolff, *Gains and Losses*, 2; Bury, *History of Freedom of Thought*, 226; Bicknell, "Unbelievers," 474, 484; Willey, *Christianity Past and Present*, 109; Budd, *Varieties of Unbelief*, 104.

A third factor that has been stressed by scholars in their interpretation of the birth of agnosticism concerns the ethical revolt against Christian orthodoxy. See Howard R. Murphy, "Ethical Revolt Against Christian Orthodoxy in Early Victorian England," 801; Wolff, *Gains and Losses*, 2; Charles Coulston Gillispie, *Edge of Objectivity*, 349; Chadwick, *Secularization of the European Mind*, 155; Tolleson, *Relation Between Leslie Stephen's Agnosticism and Voltaire's Deism*, 54–55.

Chapter Two
Mansel and the Kantian Tradition

Epigraph from Mansel, *LRT*, xli.

1. William Whewell, *Letter to the Author of Prolegomena Logica*, 7.

2. John William Burgon, "Henry Longueville Mansel," 149. Besides this chapter on Mansel in the second volume of *Lives of Twelve Good Men*, the other important biographical source is the *Dictionary of National Biography* entry on Mansel by Leslie Stephen. Although the religious backgrounds of Mansel's biographers were quite different, both had important insights to offer on the significance of Mansel's work. John William Burgon (1813–1888) was dean of Chichester and a High Churchman of the old school. Scorned for his extreme reactionism, Burgon could sympathize with the trials Mansel endured as a result of his controversial method of defending the conservative position. But one could argue that Stephen made a more appropriate biographer in that he was better equipped than Burgon to understand the agnostic aspect of Mansel's work.

3. B. A. Knox, "Filling the Oxford Chair of Ecclesiastical History, 1866," 62–70.

4. A small but respectable body of scholarly literature has grown up around Mansel. The major works are Don Cupitt, "Mansel and Maurice on Our

Knowledge of God," 301–11; Don Cupitt, "Mansel's Theory of Regulative Truth," 104–26; Don Cupitt, "What was Mansel Trying to Do?," 544–47; D. W. Dockrill, "Doctrine of Regulative Truth and Mansel's Intentions," 453–65; Kenneth D. Freeman, *Role of Reason in Religion*; Silvestro Marcucci, *Henry L. Mansel*; W. R. Matthews, *Religious Philosophy of Dean Mansel*; R. V. Sampson, "Limits of Religious Thought," 63–80; Hamish F. G. Swanston, "Henry Longueville Mansel," in *Ideas of Order*, 53–73. Mansel also finds his way into most studies of F. D. Maurice due to the controversy between the two men concerning Mansel's *Limits of Religious Thought* from 1859 to 1860. There is still a tendency for scholars to view Mansel mainly as an antagonist to Maurice, who is considered to be a much greater figure.

5. Mansel published a series of books in the fifties, including *Prolegomena Logica* (1851), an edition of Aldrich's *Artis Logicae Rudimenta* (1852), *Psychology the Test of Moral and Metaphysical Philosophy* (1855), *A Lecture on the Philosophy of Kant* (1856) as well as *The Limits of Religious Thought* (1858). Mansel's *Metaphysics* (1860) was originally written as an entry for the *Encyclopaedia Britannica*.

6. Chadwick, *Victorian Church*, pt. 1, 556; Anon., "Oxford Rationalism and the New Bampton Lectures," 237.

7. Burgon, "Henry Longueville Mansel," 216.

8. Rev. J. J. Lias et al., "Is It Possible to Know God?," 98, 132.

9. William Hale White, *Autobiography and Deliverance*, 14.

10. Mansel, *Letters, Lectures, and Reviews*, 189.

11. Hamilton's essay "On the Philosophy of the Unconditioned," which appeared as a review of M. Cousin's *Course of Philosophy* in volume 50 of the *Edinburgh Review* in October of 1829, was particularly influential. Hamilton's main works can be found in two sources, his *Discussions on Philosophy and Literature, Education and University Reform* and *Lectures on Metaphysics and Logic*.

12. George Elder Davie, *Democratic Intellect*, 122.

13. John Skelton, *Table-Talk of Shirley*, 41.

14. Mansel, *Limits of Demonstrative Science*, 3.

15. Gisela Shaw, *Das Problem des Dinges an sich in der englischen Kantinterpretation*, 26.

16. Reardon asserts that "any direct influence Kantian philosophy might have had on English theology, apart from Coleridge, was not extensive." (Reardon, *From Coleridge to Gore*, 12.) Merz confirms that Kant was virtually ignored by English intellectuals but sees Hamilton as the chief link between Kant and England. According to Merz, Mansel's Bampton Lectures "renewed attention to the philosophy of Kant which had so far affected English thought mainly in the interpretation of Sir William Hamilton." (Merz, *History of European Thought in the Nineteenth Century* 4:215.) Shaw, Welleck, and Hoaglund all argue that, with the exception of Coleridge, Hamilton, and Mansel, no major English thinker dealt with Kant seriously from 1800 up to the 1860s. (Shaw, *Das Problem des Dinges*; John Hoaglund, "Thing in Itself in English Interpretations of Kant," 1–14; René Welleck, *Immanuel Kant in England, 1793–1838*.) Carré points out that only a few slight commentaries on

Kant's philosophy appeared in England by 1850, no translation of *The Critique of Pure Reason* was attempted before 1838, and a comprehensive investigation of Kant's system was presented only after 1860. (Meyrick H. Carré, *Phases of Thought in England*, 359–60.)

17. Reardon, *From Coleridge to Gore*, 250.

18. Mansel, *Limits of Demonstrative Science*, 2.

19. Mansel, *LPK*, 5; Mansel, *LRT*, xliii–xliv; Mansel, *PC*, 66–68.

20. Norman Hampson, *Enlightenment*, 127.

21. David Hume, *Treatise of Human Nature* 1:253–54.

22. Popkin, *High Road to Pyrrhonism*, 76.

23. Kant, *Prolegomena to Any Future Metaphysics*, 42.

24. Mansel, *Metaphysics*, 312.

25. Hamilton, *LML* 1:397; Hamilton, *DPL*, 51; S. V. Rasmussen, *Philosophy of Sir William Hamilton*, 68.

26. Mansel, *PL*, 296; Mansel, *Letters, Lecturers, and Reviews*, 193.

27. Mansel's method of overcoming idealism was to posit the existence of a faculty of intuition which, unlike the senses, could inform us of the existence of objects other than our own nervous systems. The "locomotive" faculty, the means by which we consciously exert ourselves through an act of volition, informs "us immediately of the existence and properties of a material world exterior to our organism. This exterior world manifests itself in the form of *something resisting our volition.*" (See ibid., 88–89.) An object is presented as transcendentally real when it resists the effort of the locomotive faculty. But Mansel's approach was really no solution to the problem. By remaining on the empirical plane, he was unable to get outside the world of his own mind. The locomotive faculty, despite its active nature, could still only yield sensations (it was, after all, only presentative), and Mansel had admitted that all sensation was only an affection of the nervous system. See also Freeman, *Role of Reason in Religion*, 24.

28. Mansel, *Metaphysics*, 221, 226; Mansel, *PL*, 97–98.

29. Hamilton, *DPL*, 16; Mansel, *Metaphysics*, 172.

30. Mansel, *Metaphysics*, 237.

31. Kant, *Groundwork of the Metaphysic of Morals*, 126.

32. Chadwick, *Victorian Church*, pt. 1, 556.

33. Popkin, *History of Scepticism*, xiv.

34. Ibid., xv, 84; Pierre Bayle, *Historical and Critical Dictionary*, xxii–xxxiii; R. R. Palmer, *Catholics and Unbelievers in Eighteenth Century France*, 219.

35. Benn, *History of English Rationalism* 1:37.

36. One book of sermons, Newman's *Fifteen Sermons Preached Before the University of Oxford* (1843), dealt especially with the theme of faith's relation to reason.

37. Richard Popkin, "Skepticism in Modern Thought," 249; S. A. Matczak, "Fideism," 909. MacQuarrie adds to the list Bonhoeffer and O. Cullmann. See John MacQuarrie, *Twentieth-Century Religious Thought*, 319.

38. Mansel, *Examination of the Rev. F. D. Maurice's Strictures on the Bampton Lectures of 1858*, 9.

39. Mansel, *PC*, 51; Hamilton, *LML* 2:374.

40. Mansel, *PL*, xi; Mansel, *LRT*, 131; Mansel, *PL*, 246; Mansel, *LPK*, 29–31.

41. Mansel, "On Miracles as Evidences of Christianity," *Aids to Faith*, 34.

42. Richard Yeo, "William Whewell, Natural Theology and the Philosophy of Science in Mid Nineteenth Century Britain," 513; Dean, "Through Science to Despair," in Paradis and Postlewait, eds., *Victorian Science and Victorian Values*, 124. Cupitt suggests that Mansel's contemporaries were hostile to his religious thought because of the popularity of the design argument. See Cupitt, "Mansel's Theory of Regulative Truth," 108. But Mansel believed that he had presented a form of natural theology based upon the idea of the mind as designed: "If it be thought no unworthy occupation for the Christian preacher to point out the evidences of God's Providence in the constitution of the sensible world and the mechanism of the human body; or to dwell on the analogies which may be traced between the scheme of revelation and the course of nature; it is but a part of the same argument to pursue the inquiry with regard to the structure and laws of the human mind. The path may be one which, of late years at least, has been less frequently trodden . . . and the lesson of the whole, if read aright, will be but to teach us that in mind, no less than in body, we are fearfully and wonderfully made by Him whose praise both alike declare." (*LRT*, 21–22.) Yet it still holds that Mansel's natural theology of the mind was so designed to destroy Paley's natural theology of nature.

43. Mansel, *LRT*, 93. Kant also used the terms "regulative" and "speculative" in his *Critique of Pure Reason*. For Kant it was valid for the ideas of pure reason, including God, to be used regulatively by reason to posit a goal and thereby to provide unity for the understanding. Mansel opposed this notion because it gave reason a positive role. Mansel admitted that his use of the term "regulative truth" was "suggested by the language of Kant," but he insisted that he used it "in a different manner from that in which Kant employs it." (Mansel, *Second Letter to Professor Goldwin Smith*, 63.) Mansel recognized that in rejecting Kant's notion of reason he had transformed Kant's notion of the regulative use of the ideas of pure reason. "As I do not adopt Kant's distinction between the understanding and the reason," Mansel affirmed, "I could not adopt his distinction between the speculative and regulative use of the latter faculty. Accordingly, I have from the first applied the distinction to the understanding alone; an application which in effect becomes the direct reverse of Kant's." (Ibid.) Once again we have a case where common concepts and terminology seem to point to a basis of agreement between Kant and Mansel but actually conceal a profound disparity. Whereas Kant viewed the transcendental idea of God in the "Transcendental Dialectic" as a legitimate principle of interpretation of the natural world when used regulatively, Mansel believed that the regulative idea of God is a finite form under which we can think of an infinite God in order to guide our actions, but he believed it was totally empty for speculative purposes.

44. Mansel, *Metaphysics*, 344.

45. Kant, *Groundwork of the Metaphysic of Morals*, 76; Mansel, *LPK*, 31; Mansel, *Metaphysics*, 332.
46. Mansel, *PL*, 128; Mansel, *Metaphysics*, 156; Hamilton, *LML* 1:138.
47. Flint, *Agnosticism*, 260, 370–71.

Chapter Three
Herbert Spencer and the Worship of the Unknowable

Epigraph from Friedrich Nietzsche, *On The Genealogy of Morals*, trans. Walter Kaufmann and R. J. Hollingdale (New York: Vintage Books, 1967), 156.
1. [J. B. Mozley], "Mansel's Bampton Lectures" (1859), 352–90; idem, "Mr. Mansel and Mr. Maurice" (1860), 283–312; Frederick Denison Maurice's *What Is Revelation?* (1859) and *Sequel to the Inquiry, What Is Revelation?* (1860); Goldwin Smith's *Rational Religion, and the Rationalistic Objections of the Bampton Lectures for 1858* (1861); [James McCosh], "Intuitionalism and the Limits of Religious Thought" (1859), 137–59; Herbert Spencer's *First Principles of a New System of Philosophy* (1862); and John Stuart Mill's *Examination of Sir William Hamilton's Philosophy* (1865). For a fuller listing of contemporary books and articles and a discussion of the Mansel controversy see Bernard Lightman, *Henry Longueville Mansel and the Genesis of Victorian Agnosticism* (1979), 160–228, 476–80.
2. Maurice, *What Is Revelation?*, 141; Anon., "English Theological Literature in 1859," 451.
3. James Martineau, *Essays, Reviews, and Addresses*, 196.
4. Lias, "Is It Possible to Know God?," 99. For a similar statement from a spokesman for unbelief see Richard Bithell, "On 'Knowing God,'" *Agnostic Annual* (1895), 65.
5. [Richard Simpson], "Mansel's Bampton Lectures," 409–10; Josef L. Altholz and Damian McElrath, eds., *Correspondence of Lord Acton and Richard Simpson* 1:108; Charles Stephen Dessain, ed., *Letters and Diaries of John Henry Newman*, 77; Mansel, *LRT*, 3rd ed., 32.
6. J. Derek Holmes, ed., *Theological Papers of John Henry Newman on Faith and Certainty*, 159–60. See also Dessain, ed., *Letters and Diaries*, 256.
7. Matczak, "Fideism," 909.
8. J. B. Mozley, *Letters of the Rev. J. B. Mozley*, 240; Mozley, "Mansel's Bampton Lectures," 375.
9. Cupitt, "Mansel and Maurice on Our Knowledge of God," 301–11; Arthur Michael Ramsay, "Maurice and Mansel," in *F. D. Maurice and the Conflicts of Modern Theology*, 72–81; Sampson, "Limits of Religious Thought," 63–80.
10. Maurice, *What is Revelation?*, 389. For more on Maurice and Mansel see Bernard Lightman, "Broad Church Reactions to the Mansel Controversy," 9–22.
11. Mansel, *Limits of Religious Thought*, 5th ed., xxxviii.
12. Mansel, *Gnostic Heresies of the First and Second Centuries*, 274.
13. Goldwin Smith, *Rational Religion*, x.

14. Mansel, *Second Letter to Professor Goldwin Smith*, 53.

15. "Coryphaeus of Agnosticism," 458; Alfred Caldecott and H. R. Mackintosh, eds., "Agnosticism," 362; Storr, *Development of English Theology*, 422.

16. James Ward, *Naturalism and Agnosticism*, 558; Seth, *Scottish Philosophy*, 178; Webb, *Study of Religious Thought*, 92; Robertson, *History of Freethought in the Nineteenth Century* 1:213; Metz, *A Hundred Years of British Philosophy*, 40; L. E. Elliott-Binns, *English Thought, 1860–1900*, 325; Hoaglund, "Thing in Itself in English Interpretations of Kant," 3; Edwyn Bevan, "Mansel and Pragmatism," 318; Ninian Smart, "Mansel on the Limits of Religious Thought," 377; Sheldon, *Unbelief in the Nineteenth Century*, 101; Dockrill, "Origin and Development of Nineteenth Century English Agnosticism," 20; J.D.Y. Peel, *Herbert Spencer*, 128; Flint, *Agnosticism*, 569.

17. Young, *Darwin's Metaphor*, 184; James R. Moore, *Post-Darwinian Controversies*, 153–55.

18. J. A. Lauwerys, "Herbert Spencer and the Scientific Movement," 173.

19. David Wiltshire, *Social and Political Thought of Herbert Spencer*, 3.

20. James R. Moore, "Herbert Spencer's Henchmen," 76–100; Vilhelm Grønbech, *Religious Currents in the Nineteenth Century*, 137.

21. Wiltshire, *Social and Political Thought of Herbert Spencer*, 5.

22. Ibid., 58; Herbert Spencer, *Autobiography* 2:19.

23. Beatrice Webb, *My Apprenticeship*, 30.

24. Wiltshire, *Social and Political Thought of Herbert Spencer*, 96.

25. Ethel F. Fiske, ed., *Letters of John Fiske*, 297.

26. Duncan, ed., *LLHS*, 80; Spencer, *Autobiography* 1:171–73.

27. [John Frederick William Herschel], "Whewell on Inductive Sciences," 181.

28. Ralph M. Blake et al., *Theories of Scientific Method*, 180; Joseph J. Kockelmans, ed., *Philosophy of Science*, 29; George Basalla et al., eds., *Victorian Science*, 400.

29. Basalla et al., *Victorian Science*, 413.

30. Spencer, *Autobiography* 1:467. Huxley once told Beatrice Webb in 1887 that Spencer "elaborated his theory from his inner consciousness." "He never reads," Huxley complained, he "merely picks up what will help him to illustrate his theories." Beatrice Webb, *My Apprenticeship*, 28.

31. Spencer, "Mill versus Hamilton," 399.

32. Spencer, *PP*, 52; Spencer, *Autobiography* 1:291.

33. Waddington, *Development of British Thought*, 129.

34. Flint, *Agnosticism*, 580; Henry Sidgwick, *Lectures on the Philosophy of Kant*, 1.

35. Herschel, "Whewell on Inductive Sciences," 206; E. W. Strong, "William Whewell and John Stuart Mill," 219; Herbert Dingle, "Scientific Outlook in 1851 and in 1951," 86.

36. J. W. Burrow, "Herbert Spencer," 679.

37. Sheldon, *Unbelief in the Nineteenth Century*, 44; Alan Hart, *Synthetic Epistemology of Herbert Spencer*, 47.

38. Spencer, "Mill versus Hamilton," 391.

39. London, University of London Library, Herbert Spencer Papers, MS 791/355/6, p. 25.

40. Wiltshire, *Social and Political Thought of Herbert Spencer*, 59. Spencer planned to send his subscribers four quarterly installments, eighty to ninety pages each, until the series was complete. As it turned out, the total number of subscribers reached 440, but when the cost of printing and binding was taken into account this left Spencer with only £120 a year. Luckily for Spencer, two hundred more transatlantic subscriptions were brought in by Edward Youmans.

41. London, University of London Library, Herbert Spencer Papers, MS 791/355/6, p. 25. The prospectus can be found in Spencer, *Autobiography* 2:557–63. An actual copy still exists at ICST-HP 7:106–7.

42. Hart, *Synthetic Epistemology of Herbert Spencer*, 106.

43. London, University of London Library, Herbert Spencer Papers, MS 791/355/6, p. 26.

44. Spencer, "Retrogressive Religion," 6–7.

45. Garvie, "Agnosticism," 219.

46. Sheldon, *Unbelief in the Nineteenth Century*, 114.

47. Cockshut, *Unbelievers*, 81.

48. Frederick Copleston, S. J., "Herbert Spencer: Progress and Freedom," in Harman Grisewood et al., *Ideas and Beliefs of the Victorians*, 89; Metz, *A Hundred Years of British Philosophy*, 105; Sheldon, *Unbelief in the Nineteenth Century*, 107–8.

49. "Coryphaeus of Agnosticism," 470; Flint, *Agnosticism*, 574.

50. Spencer, *Autobiography* 2:86.

51. Spencer, "Religion," 12.

52. Moore, "Herbert Spencer's Henchmen," 81; John Fiske, *Excursions of an Evolutionist*, 296, 301.

53. Ethel Fiske, ed., *Letters of John Fiske*, 479.

54. Webb, *My Apprenticeship*, 27.

Chapter Four
Disillusionment with and Attack on Orthodoxy

Epigraph from Stephen, *FP*, 116.

1. Frank Turner, "Victorian Conflict between Science and Religion," 356–76.

2. C. E. Plumptre, "On Professor Tyndall's Comparative Eclipse of Fame," *Agnostic Annual* (1903), 34; Chadwick, *Victorian Church*, pt. 2, 12–13; Turner, "John Tyndall and Victorian Scientific Naturalism," in Brock et al., eds., *John Tyndall*, 170.

3. Although Stephen can be credited with helping to popularize the term, Maitland, Benn, and Tolleson overestimate his contribution. Stephen was rarely mentioned by his contemporaries or other agnostics as being of decisive significance for the development of agnosticism. Maitland, *LLLS*, 281–82;

Benn, *History of English Rationalism* 2:383–84; Tolleson, *Relation Between Leslie Stephen's Agnosticism and Voltaire's Deism*, 59.

4. William James, *Essays in Pragmatism*, 92.

5. J. G. Crowther, "John Tyndall," 173.

6. Herbert Spencer, "The Late Professor Tyndall," 401; Maynard Shipley, "Forty Years of a Scientific Friendship," 253.

7. Roy M. MacLeod, "X-Club," 305–22; J. Vernon Jensen, "Interrelationships within the Victorian 'X Club,'" 539–52; Ruth Barton, *X Club.*

8. RI-TP, Typescript Bound Journals of T. A. Hirst 4:1702.

9. Barton, *X Club*, 19.

10. In his *Mausoleum Book* Stephen called Huxley a good friend and said that he appreciated Huxley's affectionate message when Stephen's second wife died. (Leslie Stephen, *Sir Leslie Stephen's Mausoleum Book*, 8.) Stephen, in a review of Leonard Huxley's biography of his father Thomas, referred to Huxley as a grand "specimen of the fighting qualities upon which Englishmen are supposed to pride themselves." See Leslie Stephen, "Thomas Henry Huxley," 3:189.

11. Stephen, *Sir Leslie Stephen's Mausoleum Book*, 100.

12. RI-TP, Tyndall Correspondence, 1784. (R.I. MSS T., 21/E6, 9.)

13. Stephen, *Some Early Impressions*, 191.

14. Moncure Daniel Conway, *Autobiography* 2:387–90.

15. ICST-HP 27:53. (Stephen to Huxley, 17 Jan., 1879.)

16. Conway, *Autobiography* 2:397. See also Wilfrid Ward, *Problems and Persons*, 255.

17. Cambridge, Cambridge University Library, Maitland Papers, Add 7001 F. W. Maitland—Life of Stephen, Mrs. W. K. Clifford's Recollections, Add 7001.23, 2. See also Maitland, *LLLS*, 336.

18. Leonard Huxley, ed., *LLTHH* 1:10–11; Bruce Gordon Murphy, *Thomas Huxley and His New Reformation*, 17.

19. Huxley, "Mr. Balfour's Attack on Agnosticism," 533–34.

20. Julian Huxley, ed., *T. H. Huxley's Diary of the Voyage of H.M.S. Rattlesnake*, 278. Inscribed by Huxley on the inside of the corner of the notebook containing the diary of his *Rattlesnake* adventures is the following quote: "*Thätige Skepsis*" "An *Active Scepticism* is that which unceasingly strives to overcome itself and by well directed Research to attain to a kind of Conditional Certainty."

21. Ibid., 26.

22. Houston Peterson, *Huxley*, 315.

23. William Irvine, *Apes, Angels, and Victorians*, 37. Irvine points out that Huxley's essay "On the Educational Value of the Natural History Sciences" (1854) reveals Huxley as a theist. The same can be said of "On Natural History, as Knowledge, Discipline, and Power" (1856). See Huxley, *SE*, 65; Michael Foster and E. Ray Lankester, eds., *Scientific Memoirs of Thomas Henry Huxley* 1:307, 311; Cyril Bibby, *T. H. Huxley*, 57.

24. William T. Jeans, *Lives of the Electricians*, 9.

25. Queenwood was an innovative school that had connections with the

Quakers and socialist Robert Owen. It offered courses in natural science, surveying, and agriculture.

26. RI-TP, Journals of John Tyndall, 196, 216, 220.

27. Cambridge, Cambridge University Library, Maitland Papers, Add 7008 F. W. Maitland Correspondence 1905–8, Caroline E. Stephen to Maitland, 8 Jan. 1905, 7008.305.

28. Noel Annan, *Leslie Stephen*, 46.

29. Stephen's daughter, Virginia Woolf, wrote in 1940 that her father "shed his Christianity—with such anguish, Fred Maitland once hinted to me, that he thought of suicide." (Virginia Woolf, *Moments of Being*, 108.) Maitland learned of Stephen's suicidal frame of mind from Sedley Taylor (1834–1920), author of books on science and music, who was at Trinity when Stephen was agonizing over his religious beliefs. In 1904 Taylor wrote a long letter to Maitland for use in the biography of Stephen. Taylor claimed that Fawcett had told him that late one night, when Fawcett had been discussing with Stephen whether or not he should resign his position as a clergyman due to his doubts, Stephen's state of mind was such that Fawcett seriously feared he might cut his throat during the night. (Cambridge, Cambridge University Library, Maitland Papers, Add 7007 F. W. Maitland Correspondence 1903–4, Add 7007.296.) Taylor went on to say that such a revelation might be too private for publication, and Maitland must have agreed, for no trace of the letter is to be found in *The Life and Letters of Leslie Stephen*. Instead, Maitland quoted the section from Stephen's *Some Early Impressions* (70) on the ease with which Stephen sloughed off the old creed, and he placed it beside a passage from a letter from an unnamed correspondent who remembered the "mental torture" Stephen experienced. (Maitland, *LLLS*, 145–47.) However sceptical we may be about Taylor's story, since the original source was the overly dramatic Fawcett, it must be concluded that Stephen suffered far more than he was willing to let on throughout his crisis of faith. See Jeffrey Paul von Arx, *Progress and Pessimism*, 11, 212–13.

30. Annan, *Leslie Stephen*, 43; Stephen, *Sir Leslie Stephen's Mausoleum Book*, 6; Stephen, *Social Rights and Duties* 1:15–16; Stephen, *Some Early Impressions*, 54.

31. F. Pollock, "Biographical," *LE* 1:5, 32; Richard Holt Hutton, "Clifford's 'Lectures and Essays' (1879)," in *Criticisms on Contemporary Thought and Thinkers* 1:258.

32. Francis E. Mineka and Dwight N. Lindley, eds., *Later Letters of John Stuart Mill 1849–1873* 15:662, 846, 927.

33. Ibid., 934.

34. Although I have elsewhere referred to Mill as an agnostic, strictly speaking this is incorrect. See Bernard Lightman, "Henry Longueville Mansel and the Origins of Agnosticism," 45–64.

35. Mill, *Autobiography*, 174.

36. John Stuart Mill, *Examination of Sir William Hamilton's Philosophy*, 96, 99.

37. Spencer, "Mill versus Hamilton," 383–413.

38. Mineka and Lindley, *Later Letters of John Stuart Mill 1849–1873* 16:1090.

39. Dallas Victor Lie Ouren, *HaMILLton*, 2; Benn, *History of English Rationalism* 2:112.

40. Dr. Gisela Shaw, "(Review) Ulke, Karl-Dieter," 132.

41. Shipley, "Forty Years of a Scientific Friendship," 253; Eve and Creasey, *LWJT*, 89.

42. London, University of London Library, Herbert Spencer Papers, MS. 791/321.

43. RI-TP, Correspondence Between Thomas Archer Hirst and John Tyndall, 622. (R.I. MSS T., 31/E7, 374.)

44. Maitland, *LLLS*, 403; Stephen, *English Utilitarians* 2:377; Stephen, *AA*, 1, 9–10; Leslie Stephen, "Mr. Maurice's Theology," 617.

45. Stephen, *AA*, 66, 94. Other examples of a specific use of the notion of limits of knowledge, mind, or thought are: Stephen, *AA*, 129; Stephen, "The Will to Believe," *Agnostic Annual* (1898), 21; Stephen, *HETEC* 1:119.

46. Stephen, *Science of Ethics*, 2; Stephen, *AA*, 144. Other uses of the term *antinomy* are: Ibid., 19, 78, 135; Stephen, *HETEC* 1:315; Stephen, "Ascendency of the Future," 801.

47. Besides using the notion of the limits of knowledge or the idea of antinomies, Stephen also buttressed his attack on aggressive theologians by offering parallel points on the restriction of reason to its proper place couched in slightly different language. He stated that some metaphysical enquiries "lie beyond the legitimate sphere of reason" (Stephen, *HETEC* 1:20), that the doctrine of the Trinity is composed of "a number of obscure statements about matters altogether above our understanding" (Stephen, *FP*, 30), and that the assumptions of theology land us in "inextricable labyrinths of dialectics" (Stephen, *AA*, 77).

48. Stephen, *English Utilitarians* 3:432; Stephen, "Ascendency of the Future," 806; Mill, *Examination of Sir William Hamilton's Philosophy*, xvii. Quoted from marginalia in the copy of Hamilton's *Discussions*, 3d ed. (Edinburgh and London: Blackwood, 1866), in the London Library, 38.

49. Stephen, *English Utilitarians* 3:382; Stephen, *Social Rights and Duties* 1:33.

50. RI-TP, Journals of John Tyndall, 414, 418.

51. Ibid., Correspondence Between Thomas Archer Hirst and John Tyndall, 22. (R.I. MSS T., 31/B4, 16.)

52. Ibid., Journals of John Tyndall, 466. In the Tyndall Papers there is a long set of notes on Kant which demonstrates Tyndall's familiarity with Kant's work. Included are a breakdown and summary of the first two critiques. Since the date of these notes is unknown it is impossible to tell if they were made in 1849 or later. See RI-TP, Sundry Manuscripts of John Tyndall, "John Tyndall. Poetry, Carlyle Etc., Philosophy Etc., Politics," Critical and Moral Philosophy of Kant. (R.I. MSS T., 3/E11, 96.)

53. Ibid., John Tyndall's MS. Note Books, "Scientific. Alpine. 8) Fragment: reminiscences of his early years in Germany," 24 (R.I. MSS T., 2/E1, 28.); Tyndall, "Scientific Use of the Imagination," in *FS* 2:101–34.

54. RI-TP, Correspondence Between Thomas Archer Hirst and John Tyndall, 414. (R.I. MSS T. 31/D3, 189.); John Tyndall, "Professor Huxley's Doctrine," 188.

55. Tyndall, *FS* 2:73; RI-TP, John Tyndall's MS. Note Books, "Religion, Carlyle, Political, Etc."(R.I. MSS T., 2/E8, 35.); Tyndall, *FS* 2:134.

56. Clifford, *LE* 1:15; Conway, *Autobiography* 2:354; Clifford, *LE* 2:247.

57. Stephen, *AA*, 127; Clifford, *LE* 1:153, 229.

58. Schurman, *Agnosticism and Religion*, 25.

59. ICST-HP 8:92. (Tyndall to Huxley, 24 Dec., [1871].)

60. Stephen, *FP*, 153–54. Stephen disliked all attempts to liberalize Christianity, since he saw liberalism and Christianity as opposites. (See Willey, "Honest Doubt," 126.) Much of his polemic against Christianity depended on this prior assumption as to what is truly Christian. Since there were really only two positions that could be held consistently, the agnostics saw Catholicism as their main adversary. Liberal Christianity, they felt, would inevitably work its way around to their position. See Moore, *Post-Darwinian Controversies*, 63–65; John Tyndall, "Note," in Andrew Dickson White, *Warfare of Science*, iv; Huxley, *SE*, 120; Clifford, *LE* 2:231.

61. Stephen, *Social Rights and Duties* 1:15; Stephen, *AA*, 52, 67.

62. Stephen, *HETEC* 1:114; Stephen, *AA*, 33. Stephen claimed Butler for agnosticism in a controversy with Gladstone. See Stephen, "Bishop Butler's Apologist," 106–22. Huxley once told Wilfrid Ward, "Butler was really one of us." See Ward, *Problems and Persons*, 251.

Chapter Five
Religion, Theology, and the Church Agnostic

Epigraph from John Morley, *On Compromise*, 153.

1. "The Agnostic Temple," *Agnostic Annual* (1885), 54. See also "Notes and Scraps," *Agnostic* 1 (January 1885), 48.

2. Engels, *Socialism Utopian and Scientific*, 13; Lenin, *Materialism and Empirio-criticism*, 347.

3. Young, *Darwin's Metaphor*, 191.

4. Elie Halévy, *England in 1815*, 424–25.

5. Jacyna draws on both in his "Scientific Naturalism in Victorian Britain," and Moore, although underlining Young's views in his rejection of the military metaphor in *Post-Darwinian Controversies* (13), has also employed Turner's perspective. (James R. Moore, "Charles Darwin Lies in Westminster Abbey," 97–113.) Even Turner has pointed to the continuity between scientific naturalism and natural supernaturalism (Frank Turner, "Victorian Scientific Naturalism and Thomas Carlyle," 325–43) and Young has discussed the changes wrought by the fragmentation of the common intellectual context provided by natural theology. (Robert Young, "Natural Theology, Victorian Periodicals and the Fragmentation of a Common Context," in Young, *Darwin's Metaphor*, 126–63.)

6. James R. Moore, *Beliefs in Science*, 41.

7. Ethel Fiske, ed., *Letters of John Fiske*, 296.

8. W. J. Blyton, "Altered Atmosphere," 191; Altick, *Victorian People and Ideas*, 192.

9. Barry, "Professor Huxley's Creed," 160.

10. Conway, *Autobiography* 2:390; Woolf, *Moments of Being*, 40; Jeans, *Lives of the Electricians*, 5; N. D. McMillan and J. Meehan, *John Tyndall*, 115.

11. Schurman, *Agnosticism and Religion*, 11; Skelton, *Table-Talk of Shirley*, 294; Bibby, *T. H. Huxley*, 45; Barry, "Professor Huxley's Creed," 165; Baumer, *Religion and the Rise of Scepticism*, 174. See also Edward Clodd, *Thomas Henry Huxley*, 142.

12. [Richard Holt Hutton], "Pope Huxley," 136. The debate did not concern religious issues but rather Huxley's ethnological lecture on Basques, Celts, and Saxons. Hutton drew the parallel between the dogmatism of theologians and the dogmatism of Huxley's ethnological doctrines because he saw agnosticism as applicable to all realms of intellectual endeavor.

13. [R. H. Hutton], "Great Agnostic," 11.

14. Maisie Ward, *Wilfrid Wards and the Transition*, 345, 349; Ward, *Problems and Persons*, 251.

15. Ward, *Problems and Persons*, 258. According to Ward, Huxley was highly pleased with an article on him in the *Quarterly Review* for 1895 because it emphasized "that side of Huxley's teaching which was consistent with the Theistic view of life—a side so often ignored by his critics." Ibid., 252. See also Barry, "Professor Huxley's Creed," 160–88.

16. Ward, *Wilfrid Wards and the Transition*, 349.

17. Lord Ernle, "Victorian Memoirs and Memories," 224.

18. L. Huxley, ed., *LLTHH* 2:243; R. H. Hutton, "Professor Huxley," 32; RI-TP, Tyndall Correspondence, 3973. (R.I. MSS T., 19/B5, 1.); Stephen, *FP*, 358.

19. Walter White, *Journals of Walter White*, 167.

20. London, University College London Library, MS Add. 136. (Clifford to Mrs. Pollock, 28 Dec. 1869.)

21. Huxley, *SHT*, ix–x; T. H. Huxley, "On the Reception of the 'Origin of Species,' " 535; Huxley, *SE*, 397; ICST–HP 15:107; Huxley, "On the Reception of the 'Origin of Species,' " 535.

22. RI-TP, Journals of John Tyndall, 33; Ibid., Tyndall Correspondence, 402. (R.I. MSS T., 10/C3, 1.)

23. Conway, *Autobiography* 2:193.

24. John Tyndall, *Glaciers of the Alps and Mountaineering in 1861*, 225; Clifford, *LE* 2:223; Stephen, *AA*, 23–24, 32, 121–22.

25. RI-TP, Tyndall Correspondence, 1265. (R.I. MSS T., 19/F6, 25.); Ward, *Problems and Persons*, 240; Huxley, "Apologetic Irenicon," 565.

26. RI-TP, Correspondence Between Thomas Archer Hirst and John Tyndall, 24. (R.I. MSS T., 31/B5, 17.)

27. Ibid., Tyndall Correspondence, 2375. (R.I. MSS T., 23/D7,40.) Tyndall also compared the persecution of freethinkers and agnostics to the oppression of the early Christians, who were known as atheists because they did not worship the Roman gods. See Tyndall, *NF*, 5.

28. Julian Huxley, ed., *T. H. Huxley's Diary*, 26; Huxley, *SCT*, 267.

29. Huxley, *MR*, 284. Years later, in 1893, Hooker was still impressed with Huxley's conception of a national Church. (See Leonard Huxley, *Life and Letters of Sir Joseph Dalton Hooker* 2:67.) Clodd quoted approvingly from Huxley's passage on a national Church. (See Clodd, *Thomas Henry Huxley*, 151.)

30. Bibby, *T. H. Huxley*, 48; Tyndall, *NF*, 29.

31. Bicknell, "Leslie Stephen's 'English Thought in the Eighteenth Century,' " 118.

32. Stephen, "Ascendency of the Future," 802; Stephen, *AA*, 306–7.

33. Stephen, *AA*, 315, 366; Stephen, "Ascendency of the Future," 802.

34. Stephen, "Triumph of Rationalism in Religion," *Agnostic Annual and Ethical Review* (1901), 7; Stephen, *FP*, 71, 360.

35. Ayres, *Huxley*, 110; Bibby, *T. H. Huxley*, 46.

36. Bibby, *T. H. Huxley*, xxii.

37. Huxley, *SCT*, 249; Huxley, *SE*, 397.

38. Benn, *History of English Rationalism* 2:387.

39. Noel Annan, ed., *Leslie Stephen: Selected Writings in British Intellectual History*, xxi.

40. Stephen, *FP*, 337; Stephen, "Triumph of Rationalism in Religion," *Agnostic Annual and Ethical Review* (1901), 8.

41. RI-TP, Correspondence Between Thomas Archer Hirst and John Tyndall, 16. (R.I. MSS T., 31/B4, 12.); ibid., Typescript Bound Journals of T. A. Hirst, 301; ibid., Correspondence Between Thomas Archer Hirst and John Tyndall, 13. (R.I. MSS T., 31/B4, 11.)

42. Ibid., Typescript Bound Journals of T. A. Hirst, 805; Ibid., Journals of John Tyndall, 1030.

43. Ibid., Sundry Manuscripts of John Tyndall, "John Tyndall. Poetry, Carlyle Etc., Philosophy Etc., Politics," Remarks on Mr. Mivart's Paper entitled, 'The Religion of Emotion,' 1–2. (R.I. MSS T., 3/E11, 96.)

44. Ibid., Tyndall Correspondence, 407. (R.I. MSS T., 10/C3, 2.); Journals of John Tyndall, 218.

45. Huxley, "Science and Religion," 35.

46. Huxley, *SE*, 395–96. See also Ruth Barton, "Evolution," 263–65; Mandelbaum, *History, Man, and Reason*, 35.

47. T. H. Huxley, "Science and 'Church Policy,' " 821. I am indebted to Ruth Barton for pointing out the existence of this little-known Huxley article. Huxley acknowledged the piece as his in an unpublished letter to F. Dyster. See ICST-HP 15:129.

48. RI-TP, Tyndall Correspondence, 403. (R.I. MSS T., 10/C3, 1.); ibid., 406. (R.I. MSS T., 10/C3, 2.); ibid., Journals of John Tyndall, 484.

49. RI-TP, John Tyndall's MS. Note Books, "Religion, Carlyle, Political, Etc.," 15–16. (R.I. MSS T., 2/E8.) Tyndall offers a similar version, but without the crucial final phrase, in *FS* 2:374.

50. RI-TP, Journals of John Tyndall, 709, 761.

51. Mandelbaum, *History, Man, and Reason*, 29.

52. Moore, *Post-Darwinian Controversies*, 68.

53. Tyndall, "Professor Huxley's Doctrine," 188.

54. Tyndall, *NF*, 395; Tyndall, *FS* 2:52, 288, 391.

55. Stephen, *English Utilitarians* 3:419; Stephen, *AA*, 379; Clifford, *LE* 1:152.

56. Benn, *History of English Rationalism* 2:386–87; Bury, *History of Freedom of Thought*, 214.

57. Nielsen, "Agnosticism," 18–20.

58. L. Huxley, ed., *LLTHH* 1:258, 260; Cockshut, *Unbelievers*, 91–92. See also Duncan, ed., *LLHS*, 101, for Spencer's reaction to Huxley's initial approval of "The Unknowable" chapter.

59. ICST-HP 25:181. (Huxley to George Rolleston, Jan. 1866.)

60. Hutton, "Moral Significance of Atheism," 22. See also [R. H. Hutton], "Militant Agnosticism," 763; John Bernard Dalgairns, "Is God Unknowable?," 617.

61. Dockrill, "T. H. Huxley and the Meaning of 'Agnosticism,' " 469.

62. Clodd, *Thomas Henry Huxley*, 220–21.

63. Peterson, *Huxley*, 315–16. In the "Prologue" (1892) to *Science and the Christian Tradition*, Huxley cast aspersions on Spencer's whole synthetic philosophy. Comparing Spencer's "Philosophy of Evolution" to Descartes's attempt "to get at a theory of the universe by the same *a priori* road," Huxley judged any such system to be "premature." Huxley, *SCT*, 41.

64. Clodd, *Thomas Henry Huxley*, 220.

65. Clodd does not give the exact dates of Huxley's letters but merely states that they were written in 1889. (See Ibid.) However, Gould's letter of inquiry to Huxley is dated 23 December 1889, and his reply to Huxley's letter is dated 2 January 1890. (See ICST-HP 17:106, 108.)

66. Fiske, ed., *Letters of John Fiske*, 412; Ward, *Wilfrid Wards and the Transition*, 350.

67. ICST-HP 28:196–97; Huxley, *SCT*, 210; Huxley, "Apologetic Irenicon," 564.

68. L. Huxley, ed., *LLTHH* 1:433. Huxley was correct in his surmise. The article dealt primarily with Renan, Strauss, Tyndall, and Spencer as representatives of the sceptical movement in modern times. Huxley was mentioned only twice in passing. See Augustus Blauvelt, "Modern Skepticism," 424–32, 582–96, 725–39.

69. "Coryphaeus of Agnosticism," 457–70; Harrison, "Future of Agnosticism," 146. Dalgairns chooses Spencer's ideas over Huxley's as an aid to expand the agnostic position more fully. (Dalgairns, "Is God Unknowable?," 617.) During one of the meetings of the Victoria Institute in 1884 a discussion arose on agnosticism in which a member stated that "there are no books now published which are doing more mischief to the cause of religion than the books of Mr. H. Spencer." (Lias et al., "Is It Possible to Know God?," 122.) See also Flint, *Agnosticism*, 569; John Morley, *Recollections* 1:110. Scholars of the twentieth century are not agreed on the importance of Spencer. Garvie sees Spencer "as the most influential of the exponents of Agnosticism," and Benn calls him the "chief of the school." (Garvie, "Agnosticism," 1:218; Benn, *History of English Rationalism* 2:204.) But Smith and Webb see Huxley

as the true agnostic because of his rejection of the Unknowable, and they therefore dismiss Spencer's claim to leadership of the movement. (Smith, "Agnosticism," 9; Webb, *Study of Religious Thought*, 83.)

70. Clifford and Tyndall were rarely discussed in the literature of the day as agnostics. This was no doubt because they never claimed in their published work to be agnostics. Stephen's "An Agnostic's Apology" was rarely mentioned either by other agnostics or orthodox Christians, and he was viewed as a minor member of the agnostic school. Benn stated that the success of agnosticism as a party badge was owing to Stephen and that from the publication of "An Agnostics Apology" dates the "world-wide celebrity of the name agnostic." (Benn, *History of English Rationalism* 2:383–84.) Similarly, Tolleson has credited Stephen with popularizing the term and keeping it from being pejorative. (Tolleson, *Relation Between Leslie Stephen's Agnosticism and Voltaire's Deism*, 59.) However, both Benn and Tolleson exaggerate Stephen's role, as there is no evidence to support their arguments.

71. RI-TP, Correspondence Between Thomas Archer Hirst and John Tyndall, 742. (R.I. MSS T., 31/F10, 546.); Ibid., Tyndall Correspondence. (R.I. MSS T., 11/F2, 254.); Ibid., Typescript Bound Journals of T. A. Hirst, 2607.

72. Ibid., Typescript Bound Journals of T. A. Hirst, 2578.

73. As published in the *Fortnightly Review* in 1876, the essay began differently from the revised version that appeared in *An Agnostic's Apology and Other Essays* in 1893. Only in the later piece does Stephen attribute the term *agnosticism* to Huxley.

74. "Gould, Frederick James," *Who was Who 1929–1940* (London: Adam & Charles Black, 1947), 3:536.

75. F. J. Gould, *Stepping-Stones to Agnosticism*, 88, 91. Other works by Gould which are important for an understanding of his agnosticism are *The Agnostic Island* (1891), *The Life Story of a Humanist* (1923), and his numerous contributions to the *Agnostic Annual* in 1890, 1894, 1896, 1898, 1899, and 1902.

76. J. M. Wheeler, *A Biographical Dictionary of Freethinkers of All Ages and Nations* (London: Progressive Publishing Company, 1889), 44.

77. Richard Bithell, *Creed of a Modern Agnostic*, 18. Among his other important works are *The Worship of the Unknowable, Agnostic Problems* (1887), and *Handbook of Scientific Agnosticism* (1892).

78. R. Bithell, "Preface," in Albert Simmons, *Agnostic First Principles*, v; R. Bithell, "An Agnostic View of Theism and Monism," *Agnostic Annual* (1896), 46; Richard Bithell, *Agnostic Problems*, 122. As an appropriate ending to his *Handbook of Scientific Agnosticism* Bithell quoted Huxley on worshiping at the altar of the "Unknown and Unknowable." Ironically, by the time Bithell's book appeared, Huxley had already regretted using the term *Unknowable*, let alone capitalizing it. (Richard Bithell, *Handbook of Scientific Agnosticism*, 64.)

79. Laing's chief works were *Modern Science and Modern Thought* (1885), *A Modern Zoroastrian* (1887), *Problems of the Future, and Other Essays* (1889), and *Human Origins* (1892).

80. Robertson, *History of Freethought in the Nineteenth Century* 2:426.
81. Budd, *Varieties of Unbelief*, 133.
82. Laing, *Modern Zoroastrian*, 126.
83. Laing, *Problems of the Future*, 197, 212.
84. Bithell, *Agnostic Problems*, 136.
85. Laing, *Modern Zoroastrian*, 4, 202.
86. Samuel Laing, "Agnostic's Creed," 13. For an expanded version see Samuel Laing, "Articles of the Agnostic Creed, and Reasons for Them," 1–2; 17–18.

Chapter Six
The New Natural Theology and the Holy Trinity of Agnosticism

Epigraph from Ward, *Naturalism and Agnosticism*, xii.
1. James Anthony Froude, *Thomas Carlyle* 2:126.
2. Stephen was quite aware of Carlyle's pervasive influence. "One proof of Carlyle's extraordinary power," Stephen wrote, "was the influence which he exercised upon men who differed from him diametrically upon speculative doctrines. Nobody, for example, represented the very antithesis to his doctrines more distinctly than J. S. Mill. And many, I believe, of Mill's disciples would be found to owe even more to the stimulus received from their dogmatic opponent than to the direct teaching of their more congenial master." Leslie Stephen, "Thomas Carlyle," 352–53. See also Turner, "Victorian Scientific Naturalism and Thomas Carlyle," 325–43.
3. T. H. Huxley, "On Natural History, as Knowledge, Discipline, and Power," *Scientific Memoirs of Thomas Henry Huxley* 1:311.
4. T. H. Huxley, "Nature," 11. Twenty-five years later Huxley reaffirmed his position. See T. H. Huxley, "Past and Present," 1–3. Huxley's views on nature are often interpreted as shifting sometime during the sixties or seventies and culminating in his lectures on evolution and ethics, where he stresses the amoral quality of the evolutionary process. See Paradis, *T. H. Huxley*, 3; Oma Stanley, "T. H. Huxley's Treatment of 'Nature,' " 120. But Huxley's attitude toward nature had always been ambivalent, paralleling his contradictory notions of theism.
5. Ward, *Wilfrid Wards and the Transition*, 347.
6. Stephen, *Social Rights and Duties* 1:40; 2:185.
7. A. T. Coslett, "Science and Value," 49.
8. Tyndall, *FS* 2:241; Huxley, *MR*, 165; RI-TP, Tyndall Correspondence, 2933. (R.I. MSS T., 14/C3, 41.)
9. RI-TP, Journals of John Tyndall, 500.
10. Barton, *X Club*, 99.
11. T. H. Huxley, "Reviews," 81.
12. Leslie Stephen, *Playground of Europe*, 148.
13. David Robertson, "Mid-Victorians amongst the Alps," 120; Maitland, *LLLS*, 83.
14. Frederic Harrison, "Sir Leslie Stephen," 439. Besides being a play-

ground, the Alps were, for Stephen, a ''cathedral'' and a ''sanctuary'' according to Maitland. (*LLLS*, 79.)

15. Maitland, *LLLS*, 296; Stephen, *Playground of Europe*, 212.

16. Stephen, *Playground of Europe*, 39; Maitland, *LLLS*, 296.

17. Stephen, *Playground of Europe*, 124, 180, 181, 197, 198, 200, 214.

18. Ibid., 213.

19. Robertson, ''Mid-Victorians amongst the Alps,'' 123. Stephen's satirical account of an ascent of the Rothorn indicates the kind of ridicule with which Tyndall had to contend. Here is Stephen's response to a question from an imaginary scientific enthusiast:

'And what philosophical observations did you make?' will be the inquiry of one of those fanatics who, by a reasoning process to me utterly inscrutable, have somehow irrevocably associated alpine travelling with science. To them I answer that the temperature was approximately (I had no thermometer) 212° (Fahrenheit) below freezing-point. As for ozone, if any existed in the atmosphere, it was a greater fool than I take it for. (Stephen, *Playground of Europe*, 39.)

20. Sara Norton and M. A. DeWolfe Howe, eds., *Letters of Charles Eliot Norton* 1:313.

21. Huxley, ''Reviews,'' 81.

22. Tyndall, *Glaciers of the Alps*, 183.

23. John Tyndall, *Hours of Exercise in the Alps*, 186; idem, *Glaciers of the Alps*, 239.

24. Tyndall, *Hours of Exercise in the Alps*, 156, 291–92.

25. Tyndall, *Glaciers of the Alps*, 240; RI-TP, Journals of John Tyndall, 1237. (18 August 1861.)

26. Huxley, *MR*, 60. The ''conception of the constancy of the order of Nature,'' Huxley claimed in 1876, ''has become the dominant idea of modern thought.'' (T. H. Huxley, *Lectures and Essays*, 12.) In ''Science and Culture'' (1880) Huxley treated the search for order as nothing less than the main activity of human beings. (Huxley, *SE*, 150.) Huxley practiced what he preached. His scientific work was concerned, especially at the beginning of his career, with problems of form and structural plan in morphology. Huxley was searching for a ''rational and natural system'' in invertebrate zoology. (Julian Huxley, ed., *T. H. Huxley's Diary*, 40, 42.) Paradis has made Huxley's idea of order the distinguishing feature of this thought, even more basic than his agnosticism. (Paradis, *T. H. Huxley*, 112.) According to Paradis Huxley needed some element of stability to balance his vision of flux and universal motion, otherwise the world would be dissolved into a blur of elemental activity. (ibid., 75.)

27. Young, *Darwin's Metaphor*, 240.

28. Neal C. Gillespie, *Charles Darwin and the Problem of Creation*, 54, 124, 131; Dov Ospovat, ''Darwin after Malthus,'' 214–15; idem., ''God and Natural Selection,'' 171, 188, 193. Some older studies have also discussed this point, though less systematically. See Maurice Mandelbaum, ''Darwin's Religious Views,'' 367; Stanley Edgar Hyman, *Tangled Bank*, 40–41.

29. Stephen, ''Evolution and Religious Conceptions,'' 381.

30. Tyndall was also not particularly enamored of the *Bridgewater Treatises*. In 1854 he was asked to edit Prout's treatise, and upon reading it, he pronounced the book to be lacking in both scientific depth and religious inspiration. "Certainly if no better Deity than this can be purchased for the eight thousand pounds of the Earl of Bridgewater," Tyndall wrote, "it is a dear bargain." (Eve and Creasey, *LWJT*, 56.)

31. RI-TP, Journals of John Tyndall, 413; Ibid., Correspondence Between Thomas Archer Hirst and John Tyndall, 93, 104. (R.I. MSS T., 31/B7, 32; 31/B7, 35.)

32. Tyndall, *FS* 2:65; Tyndall, *NF*, 12, 346.

33. Huxley, "On Natural History, as Knowledge, Discipline, and Power," in Foster and Lankester, eds., *Scientific Memoirs of Thomas Henry Huxley* 1:307. In order to illustrate that nature exhibits design similar to the human intellect, Huxley drew upon an analogy that he picked up mountaineering with Tyndall. In *The Glaciers of the Alps* Tyndall recounted an expedition of 1856 to the Bernese Alps in southern Switzerland which he made with Huxley:

> Once on turning a corner an exclamation of surprise burst simultaneously from my companion and myself. Before each of us and against the wall of fog, stood a spectral image of a man, of colossal dimensions; dark as a whole, but bounded by a colored outline. We stretched forth our arms; the spectre did the same. We raised our alpenstocks; the spectres also flourished their batons. All our actions were imitated by these fringed and gigantic shades. We had, in fact, the *Spirit of the Brocken* before us in perfection. (22)

The *Brocken Spectre* is a term used by mountaineers to describe the phenomenon in which the shadows of the climbers, greatly magnified, are projected on the mists about the summit of the mountain opposite. (Anthony Huxley, ed., *Standard Encyclopedia of the World's Mountains* [London: Weidenfeld and Nicolson, Educational, Ltd., 1962], 165.) The name for this unusual optical illusion is taken from the Brocken peak in the Harz Mountains in Germany either because this is where the spectre may have been first seen, or, more likely, because of the supernatural legends associated with the Brocken. When Goethe whisked Faust to the summit of the Brocken for the revels of Walpurgis Night, the mountain had already acquired its connection with German legends of witchcraft and evil. (Ibid., 165; Peter Crew, *Encyclopaedic Dictionary of Mountaineering* [London: Constable, 1968], 31.) Huxley was obviously impressed by the spectacular image that appeared in the mist, and treated the vision as divine-like. Note that Tyndall's description underscored the fact that the spectre was merely a reflection of human movements. Tyndall encountered the Spectre of the Brocken a second time in 1890. (See Tyndall, *NF*, 330.)

34. Huxley, "Science and Religion," 35.

35. T. H. Huxley, *Darwiniana*, 20, 223.

36. Francis Darwin and A. C. Seward, eds., *More Letters of Charles Darwin* 1:386. Darwin had complained of his champion earlier, in 1860, for similar reasons. See Ibid. 1:139.

37. T. H. Huxley, *SCT*, 41. Just before his death Huxley reiterated his stand on natural selection. See Huxley, "Past and Present," 3. See also Edward Bagnall Poulton, "Thomas Henry Huxley and the Theory of Natural Selection," 193; Michael Bartholomew, "Huxley's Defence of Darwin," 525–35.

Huxley's reservations about natural selection stemmed from his demand for proof in the form of the production of mutually infertile breeds (the true mark of distinct species) from a single stock by means of artificial selection. If artificial selection could not produce what Darwin claimed for natural selection, then Huxley saw no reason to tie the fortunes of a naturalistic theory of evolution to one unverified hypothesis. Huxley maintained that Darwinism should be accepted only as a "working hypothesis" and that scientists should "see what could be made of it." (Thomas Henry Huxley, "On the Reception of the 'Origin of Species,' " 551.) However, Huxley himself made little use of natural selection theory in his own work to solve biological problems. Certainly, there is no evidence of a radical change in Huxley's scientific studies after 1859, and Ghiselin has concluded that he "remained a pre-Darwinian anatomist as long as he lived." (Michael T. Ghiselin, "Individual in the Darwinian Revolution," 125.) Huxley favored supplementing natural selection with saltations or mutations. (Leonard Huxley, ed., *LLTHH* 1:189; Huxley, *Lay Sermons, Addresses, and Reviews*, 297.) It could only have shocked Darwin that Huxley was arguing for *per saltum* evolution, for it allowed thinkers such as Mivart to smuggle in miraculous creation through the back door. Darwin continually declared himself in favor of a theory of evolution which was continuous and nonsaltative.

38. Tyndall agreed with Huxley, even as late as 1870, that natural selection was a hypothesis to be proved or disproved in the future. See Tyndall, *FS* 2:133. Clifford, however, accepted natural selection without any reservations. He referred to the process of evolution as "long, cumbrous, and wasteful." See Clifford, *LE* 1:213.

39. Huxley, "On the Reception of the 'Origin of Species,' " 554.

40. Huxley, "Apologetic Irenicon," 567.

41. Stephen, *FP*, 78–79; Stephen, "Evolution and Religious Conceptions," 382–83; Stephen, *AA*, 349.

42. Karl Pearson, *Life, Letters and Labours of Francis Galton* 3A:435. For Galton's sense of the limits of human knowledge see Ibid. 3B:472.

43. Ibid. 1:207.

44. Francis Galton, "Part of Religion in Human Evolution," 758.

45. Pearson, *Life, Letters and Labours of Francis Galton* 3A:271–72.

46. ICST-HP 15:107; Stephen, *AA*, 369; Clifford, *LE* 1:71; RI-TP, Journals of John Tyndall, 383.

47. Sidgwick, *Lectures on the Philosophy of Kant*, 374.

48. Huxley, *Darwiniana*, 165; Stephen, *Science of Ethics*, 4.

49. This is one of the main themes of Turner's *Between Science and Religion*. Two fine articles on Balfour have appeared recently: John David Root, "Philosophical and Religious Thought of Arthur James Balfour," 120–41; L. S. Jacyna, "Science and Social Order in the Thought of A. J. Balfour," 11–34. Jacyna is especially sharp on examining how Balfour's epistemological criti-

cism was motivated by a desire to defend the conservative social implications he drew from science and to reject the more radical ones deduced by scientific naturalism. For a more detailed discussion of the agnostic attack on a priori knowledge, and of their inability to justify their holy trinity, see Bernard Lightman, *Henry Longueville Mansel and the Genesis of Victorian Agnosticism*, 371–434.

50. Huxley, *Hume*, 118; Leonard Huxley, ed., *LLTHH* 1:261; Stephen, "On Some Kinds of Necessary Truth," 59; Clifford, *LE* 1:277, 329.

51. Huxley, *Hume*, 85; Stephen, *AA*, 135, 147.

52. Tyndall, *FS* 2:186–87; Stephen, *HETEC* 1:56; Clifford, *LE* 1:99, 100, 278, 282.

53. Huxley, *Hume*, 65.

54. Stephen, *Social Rights and Duties* 2:205, 209; Clifford, *Common Sense of the Exact Sciences*, 226.

55. John D. North, "William Kingdon Clifford," 322.

56. London, University College London Library, MS. Add. 172, [Lecture Notes on Geometry], 1; James R. Newman, "William Kingdon Clifford," *Scientific American* 188 (Feb. 1953), 80.

57. Clifford, *Mathematical Papers*, 21; J. D. North, *Measure of the Universe*, 73.

58. J. J. Callahan, "Curvature of Space in a Finite Universe," 99.

59. Clifford, *LE* 1:281, 293; Joan L. Richards, "Reception of a Mathematical Theory," 143–66.

60. Leslie Stephen, "Belief and Evidence," 12 June 1877, Metaphysical Society Papers, 2657 e.l., The Bodleian Library, Oxford, 4; Huxley, *SCT*, 70; Tyndall, *FS* 2:342; Tyndall, "Professor Huxley's Doctrine," 188.

61. Leslie Stephen, "Philosophic Doubt," 167, 170; Huxley, *MR*, 176. See also Barton, "Evolution," 269.

62. Howard Edward Smokler, *Scientific Concepts and Philosophical Theory*, 20–21.

63. Huxley, *MR*, 60; Huxley, *SHT*, 47; Huxley, *Hume*, 127.

64. Stephen, *AA*, 82; Stephen, *Science of Ethics*, 9; Tyndall, *FS* 1:343; 2:85.

65. Huxley, *Hume*, 121.

66. Huxley, *MR*, 41; Huxley, "On the Reception of the 'Origin of Species,' " 553. See also Dockrill, "Origin and Development of Nineteenth Century English Agnosticism," 24.

67. Wladyslaw Krajewski, "Idea of Statistical Law in Nineteenth Century Science," 398, 401, 404.

68. P. M. Heimann, "Molecular Forces, Statistical Representation and Maxwell's Demon," 199; Merz, "On the Statistical View of Nature," in *A History of European Thought in the Nineteenth Century* 2:599.

69. Heimann, "Molecular Forces, Statistical Representation and Maxwell's Demon," 201.

70. Krajewski, "Idea of Statistical Law," 401; Loren Eiseley, *Darwin and the Mysterious Mr. X*, 214; François Jacob, *Logic of Life*, 166–67, 197.

71. Philip P. Wiener, *Evolution and the Founders of Pragmatism*, 83; Silvan S. Schweber, "Origin of the 'Origin' Revisited," 271.

72. Niels Bohr, *Atomic Theory and the Description of Nature*, 18; Werner Heisenberg, *Physics and Philosophy*, 58, 81.

73. David M. Knight, *Atoms and Elements*, 2; Sir Basil Schonland, *Atomists*, 25; W. H. Brock and D. M. Knight, "Atomic Debates," 24.

74. Leonard Huxley, *Life and Letters of Sir Joseph Dalton Hooker* 2:359. Stephen and Huxley believed that they could maintain an agnostic position on the existence of atoms but still use atomic theory where practical. (Huxley, *EE*, 132; Stephen, "Philosophic Doubt," 166.)

75. Tyndall, *FS* 2:108. Tyndall argued that many scientists were inconsistent, for the wave theory of light, which was generally accepted as an adequate or comprehensive explanation of the facts, implied the existence of atoms. (Ibid. 2:109.) Some of Tyndall's scientific research was aimed at demonstrating the significance of atomic theory, particularly his work on radiant heat.

76. Tyndall, *FS* 2:385; Huxley, *MR*, 75; Stephen, *AA*, 131.

77. Stephen G. Brush, "Irreversibility and Indeterminism," 616.

78. Wiener, *Evolution and the Founders of Pragmatism*, 29, 82.

79. William A. Wallace, *Causality and Scientific Explanation* 2:165.

80. Francis Darwin, ed., *Autobiography of Charles Darwin*, 66.

81. RI-TP, Tyndall Correspondence, 3022. (R.I. MSS T., 14/C7, 70.) The same letter is also located at ICST-HP 8:155.

82. Spencer, "Late Professor Tyndall," 404.

83. Ward, *Naturalism and Agnosticism*, xii; Sidgwick, *Lectures on the Philosophy of Kant*, 391.

84. Tyndall, *NF*, 365. The full song can be found in Charles Neaves, *Songs and Verses Social and Scientific*, 55–59.

85. RI-TP, Journals of John Tyndall, 1328.

86. According to Irvine, Huxley was present during this festive occasion, and he heard Neaves's song, although his letters and essays contain no mention of the event. See Irvine, *Apes, Angels, and Victorians*, 242–43.

87. Leonard Huxley, ed., *LLTHH* 2:420; Stephen, "Philosophic Doubt," 165; Stephen, *English Utilitarians* 3:407.

88. Huxley, *MR*, 61, 211; Stephen, *AA*, 148, 165.

89. Huxley, *Hume*, 48. See also Smokler, *Scientific Concepts and Philosophical Theory*, 27, 40; Clifford, *LE* 1:290.

90. Tyndall, *FS* 1:28; Huxley, "Bishop Berkeley on the Metaphysics of Sensation," *Hume* (1897), 302. See also Clifford, *Seeing and Thinking*, 69.

91. Tyndall, *FS* 2:193. See also Clifford, *LE* 2:71; Huxley, *MR*, 210.

92. Huxley, *MR*, 193–94; Huxley, *EE*, 130; Huxley, "Bishop Berkeley on the Metaphysics of Sensation," *Hume*, 279; Huxley, "On Sensation and the Unity of Structure of Sensiferous Organs," *Hume* (1897), 308; Tyndall, *FS* 2:192; Stephen, *AA*, 135, 143.

93. Huxley, "Bishop Berkeley on the Metaphysics of Sensation," *Hume*, 259.

94. Clifford, *LE* 1:294; 2:74, 84–87, 143. Clifford also justified the belief in an external world in the same way that he argued for the uniformity of nature—by saying that those who make use of it survive the struggle for existence. See Ibid. 2:74; Smokler, *Scientific Concepts and Philosophical Theory,* 49.

95. James Ward, *Naturalism and Agnosticism,* 381.

Conclusion: The Tragedy of Agnosticism

Epigraph from George Orwell, *The Collected Essays, Journalism, and Letters of George Orwell: My Country Right or Left, 1940–1942* 2:15–16.

1. Dockrill puts the terminal date at 1893. See Dockrill, "Origin and Development of Nineteenth Century English Agnosticism," 29.

2. Dockrill, "Studies in Nineteenth Century English Agnosticism," 494.

3. RI-TP, Journal 1889, 93.

4. Murphy suggests that the very frequency of Huxley's articles "would seem to indicate not that Huxley had arrived at a secure position, but that he was, to some degree, growing more uncertain." (Murphy, *Thomas Huxley and His New Reformation,* 222.)

5. Wiltshire, *Social and Political Thought of Herbert Spencer,* 100.

6. Webb, *My Apprenticeship,* 90. Lauwerys attributes the cause of Spencer's deepening pessimism to his consciousness that the second law of thermodynamics meant that the universe was running down toward death and uniformity. This may have shaken Spencer's faith in science. See Lauwerys, "Herbert Spencer and the Scientific Movement," 191.

7. Peel, *Herbert Spencer,* 31.

8. Wiltshire points to the surge of xenophobia, jingoism, and colonial expansion which overwhelmed Spencer's stress on peace and progress and to the challenge to Spencer's individualistic liberalism from the new liberalism of T. H. Green, which contained neo-Hegelian elements. See Wiltshire, *Social and Political Thought of Herbert Spencer,* 100

9. Chadwick, *Victorian Church,* pt. 2, 114.

10. Herbert Spencer, "Ultimate Questions," in *Facts and Comments,* 288–92.

11. Morley, *Recollections* 1:114.

12. Woolf, *Moments of Being,* 41. Stephen was, according to Virginia, impossible to live with. He became a "tyrant"—at times pathetically vulnerable and self-pitying but at other moments violent and exacting. "It was," she remembered, "like being shut up in the same cage with a wild beast." Ibid., 116.

13. Oxford, Bodleian Library, Bryce Papers. (Leslie Stephen to James Bryce, 20 June 1898.)

14. Comments in a speech delivered in 1892 already reveal this feeling. (Stephen, *Social Rights and Duties* 1:38.) The death of his wife no doubt heightened Stephen's sense of isolation.

15. Woolf, *Moments of Being,* 147.

16. Benn, *History of English Rationalism* 2:385.
17. F. J. Gould, *The Life-Story of a Humanist*, 165.
18. Turner, *Between Science and Religion*, 228.
19. "Perplexed Moralists," 556.
20. Dockrill, "Studies in Nineteenth Century English Agnosticism," 496.
21. Jacyna, "Scientific Naturalism and Victorian Britain," 279, 291.
22. Christine Fleming Heffner, "Incense and Salt," 15; Alan Montefiore, "Aspects of Agnosticism and Ecumenicalism," 27. The tendency of some unbelievers to put agnosticism to this use was commented upon at the turn of the century. In 1905 Armstrong asserted that Huxley detested intellectual laziness, but "his own coined and patented appellation, 'Agnostic,' was worn as a badge by myriads who had never done a hard hour's thinking in their lives, but used it for a cover for sheer intellectual laziness and contented letting alone of the most stimulating and urgent questions that can occupy the mind of man." (Armstrong, *Agnosticism and Theism in the Nineteenth Century*, 78.)
23. Maurice, *What is Revelation?*, 331.
24. Arthur Maltby, "Agnosticism," 32. The logical positivists can be placed here, and a number of commentators have viewed them as the heirs to the agnostic tradition. See Holloway, "Agnosticism," 208; Hepburn, "Agnosticism," 58; Nielsen, "Agnosticism," 24.
25. Demant has perceptively remarked on the paradoxical quality of the whole agnostic conception of our relation to nature. The agnostics said that science was helping us to subdue natural forces to human will, but they also affirmed that we are a product of nature and subject to natural law. There was always an unresolved tension between the determinism of those forces and their attempt to control nature in order to liberate humanity. (Canon V. A. Demant, "Man and Nature," *Ideas and Beliefs of the Victorians*, 236.) The same contradiction appears in agnostic efforts to construct a science of ethics.

BIBLIOGRAPHY

All works in the bibliography are cited in the footnotes. Dates in square brackets refer to first editions of primary sources.

Agnostic (A Monthly Journal of Liberal Thought). Vols. 1 and 2 (Jan.–Dec. 1885).

Agnostic Annual, 1884, 1885, 1887, 1889–91, 1893–1907. (Continued as the *Agnostic Annual and Ethical Review* 1901–7).

Altholz, Josef L., and Damian McElrath, eds. *The Correspondence of Lord Acton and Richard Simpson*. 3 vols. Cambridge: Cambridge University Press, 1971.

Altick, Richard D. *Victorian People and Ideas: A Companion for the Modern Reader of Victorian Literature*. New York: W. W. Norton & Co., 1973.

Annan, Noel Gilroy. "The Intellectual Aristocracy." In *Studies in Social History: A Tribute to G. M. Trevelyan*. Ed. J. H. Plumb. London, New York, and Toronto: Longmans, Green & Co., 1955, 243–87.

———. *Leslie Stephen: His Thought and Character in Relation to His Time*. Cambridge: Harvard University Press, 1952.

———, ed. *Leslie Stephen: Selected Writings in British Intellectual History*. Chicago and London: University of Chicago Press, 1979.

Armstrong, Richard A. *Agnosticism and Theism in the Nineteenth Century: An Historical Study of Religious Thought*. London: Philip Green, 1905.

Ayres, Clarence. *Huxley*. New York: W. W. Norton & Co., 1932.

Barker, Eileen. "Thus Spake the Scientist: A Comparative Account of the New Priesthood and Its Organisational Bases." *Annual Review of the Social Sciences of Religion* 3 (1979): 79–103.

[Barry, William.] "Professor Huxley's Creed." *Quarterly Review* 180 (Jan. 1895): 160–88.

Bartholomew, Michael. "Huxley's Defence of Darwin." *Annals of Science* 32 (1975): 525–35.

Barton, Ruth. "Evolution: The Whitworth Gun in Huxley's War for the Liberation of Science from Theology." In *The Wider Domain for Evolutionary Thought*. Ed. D. Oldroyd and I. Langham. Dordrecht, Holland; Boston; and London: D. Reidel Publishing Co., 1983, 261–87.

———. *The X Club: Science, Religion, and Social Change in Victorian England.* Ann Arbor, Mich., and London: University Microfilms International, 1980. Ph.D. diss., University of Pennsylvania, 1976.

Basalla, George, William Coleman, and Robert H. Kargon, eds. *Victorian Science: A Self-Portrait from the Presidential Addresses of the British Association for the Advancement of Science.* Garden City, N.Y.: Doubleday & Co., 1970.

Baumer, Franklin L. *Modern European Thought: Continuity and Change in Ideas, 1600–1950.* New York: Macmillan Co.; London: Collier Macmillan, 1977.

———. *Religion and the Rise of Scepticism.* New York: Harcourt, Brace & Co., 1960.

Bayle, Pierre. *Historical and Critical Dictionary: Selections.* Trans. Richard H. Popkin. Indianapolis and New York: Bobbs-Merrill Co., 1965 [1697].

Becker, Carl L. *The Heavenly City of the Eighteenth-Century Philosophers.* New Haven and London: Yale University Press, 1970.

Benn, Alfred William. *The History of English Rationalism in the Nineteenth Century.* 2 vols. London, New York, and Bombay: Longmans, Green & Co., 1906.

Bevan, Edwyn. "Mansel and Pragmatism." In *Symbolism and Belief.* Boston: Beacon Press, 1957, 318–40.

Bibby, Cyril. *T. H. Huxley: Scientist, Humanist and Educator.* New York: Horizon Press, 1960.

Bicknell, John W. "Leslie Stephen's 'English Thought in the Eighteenth Century': A Tract for the Times." *Victorian Studies* 6 (Dec. 1962): 103–20.

———. "Neologizing." *Times Literary Supplement* (29 June 1973): 749.

———. "The Unbelievers." In *Victorian Prose: A Guide to Research.* Ed. David J. DeLaura. New York: Modern Language Association of America, 1973, 469–527.

Bithell, Richard. *The Creed of a Modern Agnostic.* London: George Routledge & Sons, 1883.

———. *Agnostic Problems.* London: Williams & Norgate, 1887.

———. *A Handbook of Scientific Agnosticism.* London: Watts & Co., 1892.

Blackham, H. J., ed. *Objections to Humanism.* London: Constable, 1963.

Blake, Ralph M., Curt J. Ducasse, and Edward H. Madden. *Theories of Scientific Method: The Renaissance through the Nineteenth Century.* Ed. Edward H. Madden. Seattle and London: University of Washington Press, 1966.

Blauvelt, Augustus. "Modern Skepticism." *Scribner's Monthly* 6 (1873): 424–32, 582–96, 725–39.

Blinderman, Charles S. "John Tyndall and the Victorian New Philosophy." *Bucknell Review* 9 (March 1961): 281–90.

———. "T. H. Huxley: A Re-evaluation of His Philosophy." *Rationalist Annual* (1966): 50–62.

Blyton, W. J. "The Altered Atmosphere." *Month* 179 (1943): 187–95.

Bohr, Niels. *Atomic Theory and the Description of Nature.* Cambridge: Cambridge University Press, 1934.

Brock, W. H., and D. M. Knight. "The Atomic Debates." *Isis* 56 (1965): 5–25.

Brock, W. H., N. D. McMillan, and R. C. Mollan, eds. *John Tyndall: Essays on a Natural Philosopher.* Dublin: Royal Dublin Society, 1981.

Brush, Stephen G. "Irreversibility and Indeterminism: Fourier to Heisenberg." *Journal of the History of Ideas* 37 (1976): 603–30.

Büchner, Louis. *Force and Matter: Empirico-Philosophical Studies, Intelligibly Rendered.* Ed. J. Frederick Collingwood. London: Trübner & Co., 1870 [1855].

Budd, Susan. *Varieties of Unbelief: Atheists and Agnostics in English Society, 1850–1960.* London: Heinemann, 1977.

Burgon, John William. "Henry Longueville Mansel: The Christian Philosopher." In *Lives of Twelve Good Men.* 2 vols. 6th ed. London: John Murray, 1889, 149–237.

Burrow, J. W. "Herbert Spencer: The Philosopher of Evolution." *History Today* 8 (1958): 676–83.

Bury, J. B. *A History of Freedom of Thought.* New York: Henry Holt & Co.; London: Williams & Norgate, 1913.

Caldecott, Alfred, and H. R. Mackintosh, eds. "Agnosticism: Mansel (1820–1871)." In *Selections from the Literature of Theism.* Edinburgh: T. & T. Clark, 1904, 360–67.

Calderwood, Henry. "Agnosticism." In *A Religious Encyclopaedia or, Dictionary of Biblical, Historical, Doctrinal, and Practical Theology.* Ed. Philip Schaff. 3d ed. Toronto, London, and New York: Funk & Wagnalls Co., 1891 1:36–39.

Callahan, J. J. "The Curvature of Space in a Finite Universe." *Scientific American* 235, no. 2 (Aug. 1976): 90–100.

Cambridge. Cambridge University Library. Maitland Papers.

Carré, Meyrick H. *Phases of Thought in England.* Oxford: Clarendon Press, 1949.

Chadwick, Owen. *The Secularization of the European Mind in the Nineteenth Century.* Cambridge: Cambridge University Press, 1977.

———. *The Victorian Church.* 2 pts. London: Adam & Charles Black, 1970.

Clifford, William Kingdon. *The Common Sense of the Exact Sciences.* London: Kegan Paul, Trench & Co., 1886 [1885].

———. *Lectures and Essays.* 2 vols. Ed. Leslie Stephen and Frederick Pollock. London: Macmillan & Co., 1879.

———. *Mathematical Papers.* Ed. Robert Tucker. London: Macmillan & Co., 1882.

———. *Seeing and Thinking.* London: Macmillan & Co., 1879.

Clodd, Edward. *Grant Allen: A Memoir.* London: Grant Richards, 1900.

———. *Thomas Henry Huxley.* Edinburgh and London: William Blackwood & Sons, 1902.

Clough, Arthur Hugh. *The Poems and Prose Remains of Arthur Hugh Clough.* Edited by his wife. 2 vols. London: Macmillan & Co., 1869; reprint St. Clair Shores, Mich.: Scholarly Press, 1969.

Cockshut, A.O.J. *The Unbelievers: English Agnostic Thought, 1840–1890.* New York: New York University Press, 1966.

Conway, Moncure Daniel. *Autobiography: Memories and Experiences of Moncure Daniel Conway.* 2 vols. Boston and New York: Houghton, Mifflin & Co., 1904; reprint, New York: Da Capo Press, 1970.

Copleston, Frederick. *A History of Philosophy: Volume VIII Modern Philosophy. Part I.* Garden City, N.Y.: Image Books, 1967.

Corbishley, Thomas. "Agnosticism." *A Catholic Dictionary of Theology.* Ed. H. Francis Davis, Aidan Williams, Ivo Thomas, and Joseph Crehan. London: Thomas Nelson & Sons, 1962 1:49–52.

"The Coryphaeus of Agnosticism." *Month* 45 (1882): 457–70.

Coslett, A. T. "Science and Value: The Writings of John Tyndall." *Prose Studies* 2, no. 1 (1979): 41–57.

Crownfield, Frederick R. "Whitehead: From Agnostic to Rationalist." *Journal of Religion* 57 (1977): 376–85.

Crowther, J. G. "John Tyndall." In *Scientific Types.* Chester Springs, Pa: Dufour Editions, 1970, 157–88.

Cupitt, Don. "Mansel and Maurice on Our Knowledge of God." *Theology* 73 (July 1970): 301–11.

———. Mansel's Theory of Regulative Truth." *Journal of Theological Studies* 18 (Apr. 1967): 104–26.

———. "What Was Mansel Trying to Do?" *Journal of Theological Studies* 22 (Oct. 1971): 544–47.

Dalgairns, John Bernard. "Is God Unknowable?" *Contemporary Review* 20 (1872): 615–30.

Dampier, Sir William Cecil. *A History of Science and Its Relations with Philosophy and Religion.* Cambridge: Cambridge University Press, 1966.

Darwin, Francis, ed. *The Autobiography of Charles Darwin.* New York: Dover Publications, 1958 [1892].

Darwin, Francis, and A. C. Seward, eds. *More Letters of Charles Darwin: A Record of His Work in a Series of Hitherto Unpublished Letters.* 2 vols. London: John Murray, 1903.

Davie, George Elder. *The Democratic Intellect: Scotland and Her Universities in the Nineteenth Century.* Edinburgh: Edinburgh University Press, 1964.

Davison, W. T. "Poetic Agnosticism: Meredith and Swinburne." *London Quarterly Review* 112 (July 1909): 127–30.

Dessain, Charles Stephen, ed. *The Letters and Diaries of John Henry Newman.* Vol. 19. London: Thomas Nelson & Sons, 1969.

Dingle, Herbert. "The Scientific Outlook in 1851 and in 1951." *British Journal for the Philosophy of Science* 2 (Aug. 1951): 85–104.

Dockrill, D. W. "The Doctrine of Regulative Truth and Mansel's Intentions." *Journal of Theological Studies* 25 (1974): 453–65.

———. "The Origin and Development of Nineteenth-Century English Agnosticism." *Historical Journal* (University of Newcastle, New South Wales) 1, no. 4 (1971): 3–31.

———. "Studies in Nineteenth-Century English Agnosticism." Ph.D. diss., Australian National University, 1964.

———. "T. H. Huxley and the Meaning of 'Agnosticism.'" *Theology* 74 (1971): 461–77.

Duncan, David, ed. *Life and Letters of Herbert Spencer.* London: Methuen & Co., 1908.

E.H.M., and J. R. Thorne. "Agnosticism: Agnostic." *Notes and Queries,* 6th ser., 6 (18 Nov. 1882): 418.

Eiseley, Loren. *Darwin and the Mysterious Mr. X: New Light on the Evolutionists.* New York: E. P. Dutton, 1979.

Eisen, Sydney. "Huxley and the Positivists." *Victorian Studies* 7 (June 1964): 337–58.

Elliott-Binns, L. E. *English Thought, 1860–1900: The Theological Aspect.* London, New York, and Toronto: Longmans, Green & Co., 1956.

Engels, Frederick. *Socialism Utopian and Scientific.* Trans. Edward Aveling. New York: International Publishers, 1975.

"English Theological Literature in 1859." *Literary Churchman* 5 (16 Dec. 1859): 450–51.

Ernle, Lord. "Victorian Memoirs and Memories." *Quarterly Review* 239 (1923): 215–32.

Eve, Arthur Stewart, and C. H. Creasey. *Life and Work of John Tyndall.* London: Macmillan & Co., 1945.

Farwell, Byron. "Neologizing." *Times Literary Supplement* (31 Aug. 1973): 1002–3.

Fiske, Ethel F., ed. *The Letters of John Fiske.* New York: Macmillan Co. 1940.

Fiske, John. *Excursions of an Evolutionist.* 10th ed. Boston: Houghton, Mifflin & Co., 1889.

Flint, Robert. *Agnosticism.* Edinburgh and London: William Blackwood & Sons, 1903.

Foster, Michael and E. Ray Lankester, eds. *The Scientific Memoirs of Thomas Henry Huxley.* 4 vols. London: Macmillan & Co.; New York: D. Appleton & Co., 1898–1902. (supplementary volume 1903).

Franklin, R. L. "Religion and Religions." *Religious Studies* 10 (1974): 419–31.

Freeman, Kenneth, D. *The Role of Reason in Religion: A Study of Henry Mansel.* The Hague: Martinus Nijhoff, 1969.

Friday, James R., Roy M. MacLeod, and Philippa Shepherd. *John Tyndall, Natural Philosopher (1820–1893): Catalogue of Correspondence, Journals, and Collected Papers.* London: Mansell, 1974.

Froude, James Anthony. *Thomas Carlyle: A History of the First Forty Years of His Life. 1795–1835.* 2 vols. New York: Harper & Brothers, 1882.

Galton, Francis, "The Part of Religion in Human Evolution." *National Review* 23 (1894): 755–63.

Garvie, Alfred E. "Agnosticism." In *Encyclopedia of Religion and Ethics.* Vols. 1 and 2. Ed. James Hastings. New York: Charles Scribner's Sons, 1928, 214–20.

Ghiselin, Michael T. "The Individual in the Darwinian Revolution." *New Literary History* 3, no. 1 (Autumn 1971): 113–34.

Gilby, T., U. Voll, and P. K. Meagher. "Agnosticism." In *Encyclopedic Dictionary of Religion.* Vol. A–E. Ed. Paul Kevin Meagher, Thomas C. O'Brien, and Consuelo Maria Aherne. Washington, D.C.: Corpus Publications, 1979, 77–78.

Gillespie, Neal C. *Charles Darwin and the Problem of Creation.* Chicago and London: University of Chicago Press, 1979.

Gillispie, Charles Coulston. *The Edge of Objectivity: An Essay in the History of Scientific Ideas.* Princeton, N.J.: Princeton University Press, 1960.

Gould, F. J. *The Agnostic Island.* London: Watts & Co., 1891.

———. *The Life-Story of a Humanist.* London: Watts & Co., 1923.

———. *Stepping-Stones to Agnosticism.* London: Watts & Co., [1890].

Gregory, Frederick. *Scientific Materialism in Nineteenth Century Germany.* Dordrecht, Holland; Boston; and London: D. Reidel Publishing Co., 1977.

Grisewood, Harman et al. *Ideas and Beliefs of the Victorians: An Historic Revaluation of the Victorian Age.* New York: E. P. Dutton & Co., 1966.

Grønbech, Vilhelm. *Religious Currents in the Nineteenth Century.* Trans. P. M. Mitchell and W. D. Paden. Lawrence, Kans.: University of Kansas Press, 1964.

Grosskurth, Phyllis. *Leslie Stephen.* Harlow, Essex: Longmans, Green & Co., 1968.

Halévy, Elie. *England in 1815.* Trans. E. I. Watkin and D. A. Barker. London: Ernest Benn, 1964.

Hallam, George W. "Source of the Word 'Agnostic.'" *Modern Language Notes* 70 (1955): 265–69.

Hamilton, Sir William. *Discussions on Philosophy and Literature, Education and University Reform.* 2d ed. London: Longman, Brown, Green & Longmans; Edinburgh: Maclachlan & Stewart, 1853.

———. *Lectures on Metaphysics and Logic.* 4 vols. Ed. H. L. Mansel and John Veitch. 1859–60; reprint of 2d ed., Stuttgart-Bad Cannstatt: Friedrich Frommann Verlag, 1970.

Hampson, Norman. *The Enlightenment.* Harmondsworth, Middlesex: Penguin Books, 1976.

Harrison, Frederic. "The Future of Agnosticism." *Fortnightly Review* 51 (1889): 144–56.

———. "Sir Leslie Stephen: In Memoriam." *Cornhill Magazine* 16 (1904): 433–43.

Hart, Alan. *The Synthetic Epistemology of Herbert Spencer.* Ann Arbor, Mich., and London: University Microfilms International, 1977. Ph.D. diss., University of Pennsylvania, 1965.

Heffner, Christine Fleming. "Incense and Salt." *Christianity Today* 1 (27 May 1957): 15–16.

Heimann, P. M. "Molecular Forces, Statistical Representation and Maxwell's Demon." *Studies in History and Philosophy of Science* 1 (1970): 189–211.

Heisenberg, Werner. *Physics and Philosophy: The Revolution in Modern Science.* New York: Harper & Row, 1962.

Hepburn, Ronald. "Agnosticism." In *The Encyclopedia of Philosophy.* Vol. 1. Ed. Paul Edwards. New York: Macmillan Co.; New York: Free Press, 1972, 56–59.

Hepburn, Ronald, David Jenkins, Howard Root, Renford Bambrough, and Ninian Smart. *Religion and Humanism.* London: British Broadcasting Corporation, 1964.

[Herschel, John Frederick William.] "Whewell on Inductive Sciences." *Quarterly Review* 68 (1841): 177–238.

Hoaglund, John. "The Thing in Itself in English Interpretations of Kant." *American Philosophical Quarterly* 10 (Jan. 1973): 1–14.

Holloway, M. R. "Agnosticism." In *New Catholic Encyclopedia.* Ed. William J. McDonald. New York: McGraw-Hill Book Co., 1967 1:205–9.

Holmes, J. Derek, ed. *The Theological Papers of John Henry Newman on Faith and Certainty.* Oxford: Clarendon Press, 1976.

Houghton, Walter E. *The Victorian Frame of Mind, 1830–1870.* New Haven: Yale University Press, 1957.

Hume, David. *A Treatise of Human Nature.* 2 vols. London: Dent; New York: Dutton, 1974 [1739].

Hutton, Richard Holt. *Criticisms on Contemporary Thought and Thinkers.* 2 vols. London: Macmillan & Co., 1894.

[———.] "The Great Agnostic." *Spectator* 75 (6 July 1895): 10–11.

[———.] "Militant Agnosticism." *Spectator* 49 (17 June 1876): 763–65.

———. "The Moral Significance of Atheism." In *Theological Essays.* London: Macmillan & Co., 1888, 1–24.

[———.] "Pope Huxley." *Spectator* 43 (29 Jan. 1870): 135–36.

———. "Professor Huxley." *Forum* 20 (Sept. 1895): 23–32.

[———.] "The Theological Statute at Oxford." *Spectator* 42 (29 May 1869): 642.

Huxley, Julian et al. *Science and Religion: A Symposium.* New York: Charles Scribner's Sons, 1931.

———, ed. *T. H. Huxley's Diary of the Voyage of H.M.S. Rattlesnake.* Garden City, N.Y.: Doubleday, Doran & Co., 1936.

Huxley, Leonard. *Life and Letters of Sir Joseph Dalton Hooker.* 2 vols. London: John Murray, 1918.

———, ed. *Life and Letters of Thomas Henry Huxley.* 2 vols. New York: D. Appleton & Co., 1900.

Huxley, Thomas Henry. "An Apologetic Irenicon." *Fortnightly Review* 58 (1892): 557–71.

———. *Darwiniana.* London: Macmillan & Co., 1893.

———. *Evolution and Ethics and Other Essays.* New York: D. Appleton & Co., 1899.

———. *Hume.* New York: Harper & Brothers, 1879 [1878].

———. *Hume: With Helps to the Study of Berkeley.* London: Macmillan & Co., 1897.

———. *Lay Sermons, Addresses, and Reviews.* New York: D. Appleton & Co., 1895.

———. *Lectures and Essays.* London: Macmillan & Co., 1910.

———. *Method and Results.* New York: D. Appleton & Co., 1897.

———. "Mr. Balfour's Attack on Agnosticism." *Nineteenth Century* 37 (March 1895): 527–40.

———. "Mr. Huxley's Doctrine." *Spectator* 39 (10 Feb. 1866): 158–59.

———. "Nature: Aphorisms by Goethe." *Nature* 1 (Nov. 1869–Apr. 1870): 9–11.

———. "On the Reception of the 'Origin of Species.'" In *The Life and Letters of Charles Darwin.* 2 vols. Ed. Francis Darwin. New York: D. Appleton & Co., 1887 1:533–58.

———. "Past and Present." *Nature* 51 (Nov. 1894–Apr. 1895): 1–3.

[———.] "Reviews: The Glaciers of the Alps." *Saturday Review* 10 (21 July 1860): 81–83.

———. *Science and Christian Tradition.* London: Macmillan & Co., 1909.

[———.] "Science and 'Church Policy.'" *Reader* 4 (31 Dec. 1864): 821.

———. *Science and Culture and Other Essays.* London: Macmillan & Co., 1882.

———. *Science and Education.* New York and London: D. Appleton & Co., 1914.

———. *Science and Hebrew Tradition.* New York: D. Appleton & Co., 1898.

[———.] "Science and Religion." *Builder* 18 (1859): 35–36.

Hyman, Stanley Edgar. *The Tangled Bank: Darwin, Marx, Frazer and Freud as Imaginative Writers.* New York: Atheneum, 1962.

Irvine, William. *Apes, Angels, and Victorians.* Cleveland and New York: World Publishing Co., 1970.

Jacob, François. *The Logic of Life: A History of Heredity.* Trans. Betty E. Spillmann. New York: Pantheon Books, 1973.

Jacyna, Leon Stephen. "Science and Social Order in the Thought of A. J. Balfour." *Isis* 71 (1980): 11–34.

———. "Scientific Naturalism in Victorian Britain: An Essay in the Social History of Ideas." Ph.D. diss., University of Edinburgh, 1980.

James, William. *Essays in Pragmatism.* Ed. Alburey Castell. New York: Hafner Publishing Co., 1948.

Jeans, William T. *Lives of the Electricians: Professors Tyndall, Wheatstone, and Morse.* London: Whittaker & Co.; London: George Bell & Sons, 1887.

Jensen, J. Vernon. "Interrelationships within the Victorian 'X Club.'" *Dalhousie Review* 51 (Winter 1971–72): 539–52.

Kant, Immanuel. *Critique of Pure Reason.* Trans. Norman Kemp Smith. New York: St. Martin's Press; Toronto: Macmillan Co., 1965 [1781].

———. *Groundwork of the Metaphysic of Morals.* Trans. H. J. Paton. New York: Harper & Row, 1964 [1785].

———. *Prolegomena to Any Future Metaphysics.* Ed. Lewis W. Beck. Indianapolis and New York: Bobbs-Merrill Co., 1950 [1783].

———. *Religion within the Limits of Reason Alone.* Trans. Theodore M. Greene and Hoyt H. Hudson. New York: Harper & Row, 1960 [1793].

Kaufmann, Walter. *From Shakespeare to Existentialism.* Garden City, N.Y.: Doubleday & Co., 1960.

Knight, David M. *Atoms and Elements: A Study of Theories of Matter in England in the Nineteenth Century.* London: Hutchinson & Co., 1967.

Knox, B. A. "Filling the Oxford Chair of Ecclesiastical History, 1866: The Nomination of H. L. Mansel." *Journal of Religious History* 5 (June 1968): 62–70.

Kockelmans, Joseph J., ed. *Philosophy of Science: The Historical Background.* New York: Free Press; London: Collier-Macmillan, 1968.

Krajewski, Wladyslaw. "The Idea of Statistical Law in Nineteenth-Century Science." *Boston Studies in the Philosophy of Science* 14 (1974): 397–405.

LaCapra, Dominick. "Rethinking Intellectual History and Reading Texts." In *Modern European Intellectual History: Reappraisals and New Perspectives.* Ed. Dominick LaCapra and Stephen L. Kaplan. Ithaca and London: Cornell University Press, 1982, 47–85.

Laing, Samuel. "The Agnostic's Creed." *Pall Mall Gazette* (29 Dec. 1888): 13.

———. "Articles of the Agnostic Creed, and Reasons for Them." *Agnostic Journal and Secular Review* (5 Jan. 1889): 1–2; (12 Jan. 1889): 17–18.

———. *Human Origins.* London: Chapman and Hall, 1892.

———. *Modern Science and Modern Thought.* New York: Humboldt Publishing Co., n.d. [1885].

———. *A Modern Zoroastrian.* London: Chapman & Hall, 1898 [1887].

———. *Problems of the Future and Other Essays.* London: Chapman & Hall, 1889.

Lauwerys, J. A. "Herbert Spencer and the Scientific Movement." In *Pioneers of English Education.* Ed. A. V. Judges. London: Faber & Faber, 1952, 160–93.

Lenin, V. I. *Materialism and Empirio-criticism: Critical Comments on a Reactionary Philosophy.* Moscow: Progress, 1970 [1908].

Lias, J. J. et al. "Is It Possible to Know God?" *Journal of the Transactions of the Victoria Institute* 17 (1884): 98–141.

Lightman, Bernard. "Broad Church Reactions to the Mansel Controversy." *Victorian Studies Association Newsletter* 28 (Fall 1981): 9–22.

———. *Henry Longueville Mansel and the Genesis of Victorian Agnosticism.* Ann Arbor, Mich., and London: University Microfilms International, 1979. Ph.D. diss., Brandeis University, 1979.

———. "Henry Longueville Mansel and the Origins of Agnosticism." *History of European Ideas* 5, no. 1 (1984): 45–64.

———. "John Stuart Mill and Immanuel Kant on Nature: Idealism in Mill's 'An Examination of Sir William Hamilton's Philosophy.' " *Mill News Letter* 14, no. 2 (Summer 1979): 2–12.

———. "Pope Huxley and the Church Agnostic: The Religion of Science." *Historical Papers* (1983): 150–63.

London. Imperial College of Science and Technology. The Huxley Papers.

London. Royal Institution of Great Britain. The Tyndall Papers.

London. University College London Library. MS. Add. 136. W. K. Clifford to Mrs. Pollock, 28 Dec. 1869, and Feb. 1870.

London. University College London Library. MS. Add. 172. W. K. Clifford's lecture notes [on geometry], 187-?

London. University of London Library. Herbert Spencer Papers.

Lyell, Mrs. Charles, ed. *Life, Letters, and Journals of Sir Charles Lyell.* 2 vols. London: John Murray, 1881.

[McCosh, James.] "Intuitionalism and the Limits of Religious Thought." *North British Review* 30 (Feb. 1859): 137–59.

Macfarlane, Alexander. "William Kingdon Clifford." In *Lectures on Ten British Mathematicians of the Nineteenth Century.* New York: John Wiley & Sons; London: Chapman & Hall, 1916, 78–91.

MacLeod, Roy M. "The X-Club: A Social Network of Science in Late-Victorian England." *Notes and Records of the Royal Society of London* 24 (1970): 305–22.

McMillan, N. D., and J. Meehan. *John Tyndall: 'X'emplar of Scientific and Technological Education.* Ed. Pauric Hogan. Dublin: N.C.E.A., 1980.

MacQuarrie, John. *Twentieth-Century Religious Thought: The Frontiers of Philosophy and Theology, 1900–1960.* London: SCM Press, 1963.

Maitland, Frederic William. *The Life and Letters of Leslie Stephen.* London: Duckworth & Co., 1906.

Mallock, William H. *The New Republic; or, Culture, Faith, and Philosophy in an English Country House.* Ed. J. Max Patrick. Gainesville: University of Florida Press, 1950 [1877].

Maltby, Arthur. "Agnosticism: A Modern Form of Gnosis?" *Evangelical Quarterly* 37 (Jan.–Mar. 1965): 32–35.

Mandelbaum, Maurice. "Darwin's Religious Views." *Journal of the History of Ideas* 19 (1958): 363–78.

———. *History, Man, and Reason: A Study in Nineteenth-Century Thought.* Baltimore and London: Johns Hopkins University Press, 1977.

Mansel, Henry Longueville. *An Examination of the Rev. F. D. Maurice's Strictures on the Bampton Lectures of 1858.* London: John Murray, 1859.

———. *The Gnostic Heresies of the First and Second Centuries.* Ed. J. B. Lightfoot. London: John Murray, 1875.

———. *A Lecture on the Philosophy of Kant.* Oxford: John Henry & James Parker, 1856.

———. *Letter, Lectures, and Reviews.* Ed. Henry W. Chandler. London: John Murray, 1873.

———. *The Limits of Demonstrative Science Considered in a Letter to the Rev. William Whewell.* Oxford: William Graham, 1853.

———. *The Limits of Religious Thought Examined in Eight Lectures.* Oxford: (Printed by J. Wright for) John Murray, 1858; 2d ed. London: John Murray, 1858; 3d ed. Boston: Gould & Lincoln, 1859; reprint, New York:

AMS Press, 1973; 4th ed. London: John Murray, 1859; 5th ed. London: John Murray, 1867.

———. *Metaphysics; or, the Philosophy of Consciousness Phenomenal and Real.* New York: D. Appleton & Co., 1871 [1860].

———. "On Miracles as Evidences of Christianity." In *Aids to Faith; A Series of Theological Essays. By Several Writers. Being a Reply to "Essays and Reviews."* Ed. William Thomson. New York: D. Appleton & Co., 1863 [1861], 7–52.

———. *The Philosophy of the Conditioned.* London and New York: Alexander Strahan, 1866. (Reprint of "The Philosophy of the Conditioned: Sir William Hamilton and John Stuart Mill." *Contemporary Review* 1 [1866]: 31–49, 185–219.)

———. *Prolegomena Logica: An Inquiry into the Psychological Character of Logical Processes.* Oxford: William Graham, 1851.

———. *A Second Letter to Professor Goldwin Smith.* Oxford: Henry Hammans, 1862.

"Mansel's Bampton Lectures." *Times* (10 Jan. 1859): 10.

Marcucci, Silvestro. *Henry L. Mansel.* Florence: P. Le Monnier, 1969.

Marshall, E. "Agnosticism." *Notes and Queries,* 6th ser., 5 (24 June 1882): 489.

Martineau, James. *Essays, Reviews, and Addresses: III Theological: Philosophical.* London: Longmans, Green & Co., 1891.

———. *A Study of Religion: Sources and Contents.* 2 vols. Oxford: Clarendon Press, 1888.

Matczak, S. A. "Fideism." *New Catholic Encyclopedia.* New York: McGraw-Hill Book Co., 1967 5:908–10.

Matthews, W. R. *The Religious Philosophy of Dean Mansel.* London: Oxford University Press, 1956.

Maurice, Frederick Denison. *Sequel to the Inquiry, What is Revelation? in a Series of Letters to a Friend; Containing a Reply to Mr. Mansel's "Examination of the Rev. F. D. Maurice's Strictures on the Bampton Lectures of 1858."* Cambridge: Macmillan and Co., 1860.

———. *What Is Revelation? A Series of Sermons on the Epiphany; to Which Are Added Letters to a Student of Theology on the Bampton Lectures of Mr. Mansel.* Cambridge: Macmillan & Co., 1859.

Merz, John Theodore. *A History of European Thought in the Nineteenth Century.* 4 vols. Edinburgh and London: William Blackwood & Sons, 1907–1914.

Metz, Rudolf. *A Hundred Years of British Philosophy.* Trans. J. W. Harvey, T. E. Jessop, and Henry Sturt. Ed. J. H. Muirhead. London: George Allen & Unwin; New York: Macmillan Co., 1938.

Mill, John Stuart. *Autobiography.* Indianapolis and New York: Bobbs-Merrill Co., 1957 [1873].

———. *An Examination of Sir William Hamilton's Philosophy and of the Principal Philosophical Questions Discussed in His Writings.* Ed. J. M. Robson. Vol. 9 of *The Collected Works of John Stuart Mill.* Toronto and Buffalo: University of Toronto Press, 1979 [1865].

Mineka, Francis E., and Dwight N. Lindley, eds. *The Later Letters of John Stuart Mill 1849–1873.* Vols. 14, 15, 16, and 17 of *The Collected Works of John Stuart Mill.* Toronto and Buffalo: University of Toronto Press; London: Routledge & Kegan Paul, 1972.

Momerie, A. W. *Agnosticism: Sermons Preached in St. Peter's, Cranely Gardens, 1883–4.* Edinburgh and London: William Blackwood & Sons, 1887.

Montefiore, Alan. "Aspects of Agnosticism and Ecumenicalism." *Theology* 80 (Jan. 1977): 22–29.

Moore, James R. *Beliefs in Science: An Introduction.* Walton Hall, Milton Keynes, Eng.: Open University Press, 1981.

———. "Charles Darwin Lies in Westminster Abbey." *Biological Journal of the Linnean Society* 17 (1982): 97–113.

———. "Herbert Spencer's Henchmen: The Evolution of Protestant Liberals in Late Nineteenth-Century America." In *Darwinism and Divinity: Essays on Evolution and Religious Belief.* Ed. John R. Durant. Oxford: Basil Blackwell, 1985, 76–100.

———. *The Post-Darwinian Controversies: A Study of the Protestant Struggle to Come to Terms with Darwin in Great Britain and America, 1870–1900.* Cambridge: Cambridge University Press, 1979.

Morley, John. *On Compromise.* London: Macmillan and Co., 1903 [1874].

———. *Recollections.* 2 vols. Toronto: Macmillan Co., 1917.

Mozley, J. B. *Letters of the Rev. J. B. Mozley.* Edited by his sister. London: Rivingtons, 1885.

[———.] "Mansel's Bampton Lectures." *Christian Remembrancer* 37 (1859): 352–90.

[———.] "Mr. Mansel and Mr. Maurice." *Christian Remembrancer* 39 (1860): 283–312.

Murphy, Bruce Gordon. *Thomas Huxley and His New Reformation.* Ann Arbor, Mich., and London: University Microfilms International, 1977. Ph.D. diss., Northern Illinois University, 1973.

Murphy, Howard R. "The Ethical Revolt Against Christian Orthodoxy in Early Victorian England." *American Historical Review* 60 (1954–55): 800–817.

Murray, James A. H., ed. "Agnostic." In *A New English Dictionary on Historical Principles.* Oxford: Clarendon Press, 1884 1:186.

[Neaves, Charles.] *Songs and Verses Social and Scientific.* 5th ed. Edinburgh and London: William Blackwood & Sons, 1879.

Newman, James R. "William Kingdon Clifford." *Scientific American* 188 (Feb. 1853): 78–84.

Newman, John Henry. *Apologia Pro Vita Sua.* Ed. David J. DeLaura. New York: W. W. Norton & Co., 1968 [1864].

Nielsen, Kai. "Agnosticism." In *Dictionary of the History of Ideas.* Ed. Philip P. Wiener. New York: Charles Scribner's Sons, 1968 1:17–27.

North, John D. "William Kingdon Clifford." In *Dictionary of Scientific Biography.* Ed. Charles Coulston Gillispie. New York: Charles Scribner's Sons, 1971 3:322–23.

———. *The Measure of the Universe: A History of Modern Cosmology.* Oxford: Clarendon Press, 1965.

Norton, Sara, and M. A. DeWolfe Howe. *Letters of Charles Eliot Norton.* 2 vols. Boston and New York: Houghton Mifflin Co., 1913.

Noxon, James. "Hume's Agnosticism." In *Hume.* Ed. V. C. Chappell. Garden City, N.Y.: Doubleday & Co., 1966, 361–83.

O'Higgins, J. "Browne and King, Collins and Berkeley: Agnosticism or Anthropomorphism." *Journal of Theological Studies* 27 (1976): 88–112.

Onions, C. T. "Agnostic." *Times Literary Supplement* (10 May 1947): 225.

Orwell, George. *The Collected Essays, Journalism and Letters of George Orwell: My Country Right or Left, 1940–1942.* Vol. 2. Ed. Sonia Orwell and Ian Angus. New York: Harcourt, Brace & World, 1968.

Ospovat, Dov. "Darwin after Malthus." *Journal of the History of Biology* 12 (Fall 1979): 211–30.

———. "God and Natural Selection: The Darwinian Idea of Design." *Journal of the History of Biology* 13 (1980): 169–94.

Ouren, Dallas Victor Lie. *HaMILLton: Mill on Hamilton—A Re-examination of Sir Wm. Hamilton's Philosophy.* Ann Arbor, Mich., and London: University Microfilms International, 1977. Ph.D. diss., University of Minnesota, 1973.

Oxford. Bodleian Library. Bryce Papers. (Leslie Stephen to Bryce, 20 June, 1898.)

Oxford. Bodleian Library. Metaphysical Society Papers. 2657e. 1.

"Oxford Rationalism and the New Bampton Lectures." *Guardian* (24 Mar. 1858): 237.

Palmer, R. R. *Catholics and Unbelievers in Eighteenth Century France.* New York: Cooper Square, 1961.

Paradis, James G. *T. H. Huxley: Man's Place in Nature.* Lincoln, Nebr., and London: University of Nebraska Press, 1978.

Paradis, James G., and Thomas Postlewait, eds. *Victorian Science and Victorian Values: Literary Perspectives.* New York: New York Academy of Science, 1981.

Passmore, John. "Darwin's Impact on British Metaphysics." *Victorian Studies* 3 (Sept. 1959): 41–54.

Paulson, Ronald. *Hogarth: His Life, Art, and Times.* 2 vols. New Haven and London: Yale University Press, 1971.

Pearson, Karl. *The Life, Letters, and Labours of Francis Galton.* 3 vols. Cambridge: Cambridge University Press, 1914–30.

Peel, J.D.Y. *Herbert Spencer: The Evolution of a Sociologist.* New York: Basic Books, 1971.

"Perplexed Moralists." *Christian Commonwealth* (London) (16 May 1901): 556.

Peterson, Houston, *Huxley: Prophet of Science.* London, New York, and Toronto: Longmans, Green & Co., 1932.

Popkin, Richard H. *The High Road to Pyrrhonism.* Edited by Richard A. Watson and James E. Force. San Diego: Austin Hill Press, 1980.

———. *The History of Scepticism: From Erasmus to Descartes.* New York, Evanston, Ill., and London: Harper & Row, 1968.

———. "Skepticism in Modern Thought." In *Dictionary of the History of Ideas.* Ed. Philip P. Wiener. New York: Charles Scribner's Sons, 1973 4:240–51.

"The Popular View of Atheism." *Saturday Review* (26 June 1880): 819–20.

Poulton, Edward Bagnall. "Thomas Henry Huxley and the Theory of Natural Selection." In *Essays on Evolution 1889–1907.* Oxford: Clarendon Press, 1908, 193–219.

"The Prevalent Phase of Unbelief." *Month* 45 (1882): 153–68.

Ramsay, Arthur Michael. *F. D. Maurice and the Conflicts of Modern Theology.* Cambridge: Cambridge University Press, 1951.

Randall, John Herman, Jr. *The Career of Philosophy.* Vol. 1. *From the Middle Ages to the Enlightenment.* New York: Columbia University Press, 1962.

Rasmussen, S. V. *The Philosophy of Sir William Hamilton.* London: Hachette; Copenhagen: Levin & Munksgaard, 1925.

Reardon, Bernard M. G. *From Coleridge to Gore: A Century of Religious Thought in Britain.* London: Longman, 1971.

Richards, Joan L. "The Reception of a Mathematical Theory: Non-Euclidean Geometry in England, 1868–1883." In *Natural Order: Historical Studies of Scientific Culture.* Ed. Barry Barnes and Steven Shapin. Beverly Hills, Calif., and London: Sage Publications, 1979, 143–66.

Robertson, David. "Mid-Victorians amongst the Alps." In *Nature and the Victorian Imagination.* Ed. U. C. Knoepflmacher and G. B. Tennyson. Berkeley, Los Angeles, and London: University of California Press, 1977, 112–36.

Robertson, J. M. *A History of Freethought in the Nineteenth Century.* 2 vols. London: Dawsons of Pall Mall, 1969.

Root, John David. "The Philosophical and Religious Thought of Arthur James Balfour (1848–1930)." *Journal of British Studies* 19, no. 2 (Spring 1980): 120–41.

Ruse, Michael. *The Darwinian Revolution.* Chicago and London: University of Chicago Press, 1979.

Sampson, R. V. "The Limits of Religious Thought: The Theological Controversy." In *1859: Entering an Age of Crisis.* Ed. Philip Appleman, William A. Madden, and Michael Wolff. Bloomington: Indiana University Press, 1961, 63–80.

Schonland, Sir Basil. *The Atomists (1805–1933).* Oxford: Clarendon Press, 1968.

Schurman, Jacob Gould. *Agnosticism and Religion.* New York: Charles Scribner's Sons, 1896.

Schweber, Silvan S. "The Origin of the 'Origin' Revisited." *Journal of the History of Biology* 10 (1977): 229–316.

Seth, Andrew. *Scottish Philosophy: A Comparison of the Scottish and German Answers to Hume.* 2d ed. Edinburgh and London: William Blackwood & Sons, 1890 [1885].

Shaw, Gisela. *Das Problem des Dinges an sich in der englischen Kantinterpretation*. Bonn: H. Bouvier, 1969.

———. "[Review] Ulke, Karl-Dieter. Agnostic Thinking in Victorian Britain [Agnostisches Denken im Viktorianischen England]. Freiburg: Karl Alber Verlag, 1980; 244 pp." *Philosophy and History* 15 (1982): 132–33.

Sheldon, Henry C. *Unbelief in the Nineteenth Century: A Critical History*. London: Charles H. Kelly, n.d. [1907].

Shipley, Maynard. "Forty Years of a Scientific Friendship." *Open Court* 34 (Apr. 1920): 252–55.

Sidgwick, Henry. *Lectures on the Philosophy of Kant*. London: Macmillan & Co.; New York: Macmillan Co., 1905; New York: Kraus Reprint Co., 1968.

Simmons, Albert (Ignotus). *Agnostic First Principles: Part I of Agnosticism: A Philosophic Synthesis. Being a Critical Exposition of the Spencerian System of Thought*. London: Watts & Co., [1885].

Simon, W. M. *European Positivism in the Nineteenth Century: An Essay in Intellectual History*. Ithaca, N.Y.: Cornell University Press, 1963.

[Simpson, Richard.] "Mansel's Bampton Lectures." *Rambler* 22 (Dec. 1858): 405–15.

Singer, Charles. *Religion and Science Considered in Their Historical Relations*. London: Ernest Benn, 1928.

Skelton, John. *The Table-Talk of Shirley: Reminiscences of and Letters from Froude, Thackeray, Disraeli, Browning, Rossetti, Kingsley, Baynes, Huxley, Tyndall, and Others*. 3d ed. Edinburgh and London: William Blackwood & Sons, 1895.

Smart, Ninian, ed. "Mansel on the Limits of Religious Thought." In *Historical Selections in the Philosophy of Religion*. London: SCM Press, 1962, 361–77.

Smith, Gerald Birney. "Agnosticism." In *A Dictionary of Religion and Ethics*. Ed. Shailer Mathews and Gerald Birney Smith. New York: Macmillan Co., 1921, 9–10.

Smith Goldwin. *Rational Religion, and the Rationalistic Objections of the Bampton Lectures for 1858*. Oxford: J. L. Wheeler, 1861.

Smokler, Howard Edward. *Scientific Concepts and Philosophical Theory: An Essay in the Philosophy of W. K. Clifford*. Ann Arbor, Mich. and London: University Microfilms International, 1977. Ph.D. diss., Columbia University, 1959.

Somervell, D. C. *English Thought in the Nineteenth Century*. London: Methuen & Co., 1964.

Sorley, W. R. *A History of British Philosophy to 1900*. Cambridge: Cambridge University Press, 1965.

Spencer, Herbert. *An Autobiography*. 2 vols. New York: D. Appleton & Co., 1904.

———. *Facts and Comments*. New York: D. Appleton & Co., 1902.

———. *First Principles of a New System of Philosophy*. New York: D. Appleton & Co., 1882 [1862].

———. "The Late Professor Tyndall." *New McClures Magazine* (March 1894): 401–8.

———. "Mill versus Hamilton: The Test of Truth." In *Essays: Moral, Political, and Aesthetic.* New York: D. Appleton & Co., 1880, 383–413.

———. *The Principles of Psychology.* London: Longman, Brown, Green & Longmans, 1855.

———. "Religion: A Retrospect and Prospect." *Nineteenth Century* 15 (Jan. 1884): 1–12.

———. "Retrogressive Religion." *Nineteenth Century* 16 (July 1884): 3–26.

Stanley, Oma. "T. H. Huxley's Treatment of 'Nature.' " *Journal of the History of Ideas* 18 (1957): 120–27.

Stephen, Leslie. *An Agnostic's Apology and Other Essays.* London: Smith, Elder & Co., 1893; reprint Westmead, Farnborough, Hants: Gregg International Publishers, 1969.

———. "The Ascendency of the Future." *Nineteenth Century* 51 (1902): 795–810.

———. "Bishop Butler's Apologist." *Nineteenth Century* 39 (1896): 106–22.

———. *The English Utilitarians.* 3 vols. London: Duckworth & Co., 1900.

———. *Essays on Freethinking and Plainspeaking.* London: Longmans, Green & Co., 1873.

———. "Evolution and Religious Conceptions." In *The Nineteenth Century: A Review of Progress during the Past One Hundred Years in the Chief Departments of Human Activity.* New York and London: G. P. Putnam's Sons, 1901, 370–83.

———. *History of English Thought in the Eighteenth Century.* 2 vols. London: Smith, Elder & Co., 1876.

———. "Mr. Maurice's Theology." *Fortnightly Review* 15 (1874): 595–617.

———. "On Some Kinds of Necessary Truth." *Mind* 14 (1889): 50–65, 188–218.

———. "Philosophic Doubt." *Mind* 5 (1880): 157–81.

———. *The Playground of Europe.* Oxford: Basil Blackwell, 1936 [1871].

———. *The Science of Ethics.* London: Smith, Elder & Co., 1882.

———. *Sir Leslie Stephen's Mausoleum Book.* Introduction by Alan Bell. Oxford: Clarendon Press, 1977.

———. *Social Rights and Duties: Addresses to Ethical Societies.* 2 vols. London: Swan Sonnenschein & Co., 1896.

———. *Some Early Impressions.* London: Leonard and Virginia Woolf at the Hogarth Press, 1924 [1903].

[———.] "Thomas Carlyle." *Cornhill Magazine* 43 (1881): 349–58.

———. "Thomas Henry Huxley." In *Studies of a Biographer.* 4 vols. London: Duckworth & Co., 1902 3:188–219.

Stewart, H. L. "J. S. Mill's 'Logic': A Post-Centenary Appraisal." *University of Toronto Quarterly* 17 (July 1948): 361–71.

Storr, Vernon F. *The Development of English Theology in the Nineteenth Century, 1800–1860.* New York, Bombay, and Calcutta: Longmans, Green & Co., 1913.

Stromberg, Roland N. *An Intellectual History of Modern Europe.* 2d ed. Englewood Cliffs, N.J.: Prentice-Hall, 1975.

Strong, E. W. "William Whewell and John Stuart Mill: Their Controversy about Scientific Knowledge." *Journal of the History of Ideas* 16 (1955): 209–31.

Swanston, Hamish F. G. *Ideas of Order: Anglicans and the Renewal of Theological Method in the Middle Years of the Nineteenth Century.* Assen, The Netherlands: Van Gorcum & Co., 1974.

Tener, Robert H. "Agnostic." *Times Literary Supplement* (10 Aug. 1967): 732.

———. "Neologizing." *Times Literary Supplement* (10 Aug. 1973): 931.

———. "Neologizing." *Times Literary Supplement* (9 Nov. 1973): 1373.

———. "R. H. Hutton and 'Agnostic.'" *Notes and Queries* 11 (Nov. 1964): 429–31.

Tjoa, Hock Guan. *George Henry Lewes: A Victorian Mind.* Cambridge, Mass., and London: Harvard University Press, 1977.

Tolleson, Floyd Clyde, Jr. *The Relation Between Leslie Stephen's Agnosticism and Voltaire's Deism.* Ann Arbor, Mich., and London: University Microfilms International, 1977. Ph.D. diss., University of Washington, 1955.

Turner, Frank Miller. *Between Science and Religion: The Reaction to Scientific Naturalism in Late Victorian England.* New Haven and London: Yale University Press, 1974.

———. "Lucretius among the Victorians." *Victorian Studies* 16 (Mar. 1973): 329–48.

———. "Public Science in Britain, 1880–1919." *Isis* 71 (1980): 589–608.

———. "Rainfall, Plagues, and the Prince of Wales: A Chapter in the Conflict of Religion and Science." *Journal of British Studies* 13, no. 2 (May 1974): 46–65.

———. "The Victorian Conflict between Science and Religion: A Professional Dimension." *Isis* 69 (1978): 356–76.

———. "Victorian Scientific Naturalism and Thomas Carlyle." *Victorian Studies* 18 (1975): 325–43.

Turner, James. *Without God, Without Creed: The Origins of Unbelief in America.* Baltimore and London: Johns Hopkins University Press, 1985.

Tyndall, John. *Fragments of Science: A Series of Detached Essays, Addresses, and Reviews.* 2 vols. London: Longmans, Green & Co., 1892; reprint, Westmead, Farnborough, Hants: Gregg International, 1970.

———. *The Glaciers of the Alps and Mountaineering in 1861.* London: J. M. Dent & Co.; New York: E. P. Dutton & Co., 1906 [1860 and 1861].

———. *Hours of Exercise in the Alps.* London: Longmans, Green & Co., 1871.

———. *New Fragments.* New York: D. Appleton & Co., 1896 [1892].

———. "Professor Huxley's Doctrine." *Spectator* 39 (17 Feb. 1866): 188.

Ulke, Karl-Dieter. *Agnostisches Denken im Viktorianischen England.* Freiburg: Karl Alber Verlag, 1980.

V.H.I.L.I.C.I.V. "Agnosticism." *Notes and Queries*, 6th ser., 6 (8 July 1882): 34.

von Arx, Jeffrey Paul. *Progress and Pessimism*. Cambridge: Harvard University Press, 1985.

Wace, Henry. *Christianity and Agnosticism: Reviews of Some Recent Attacks on the Christian Faith*. Edinburgh and London: William Blackwood & Sons, 1895.

Waddington, M. M. *The Development of British Thought from 1820 to 1890: With Special Reference to German Influences*. Toronto: J. M. Dent & Sons, 1919.

Wallace, William A. *Causality and Scientific Explanation*. 2 vols. Ann Arbor, Mich.: University of Michigan Press, 1974.

Ward, James. *Naturalism and Agnosticism: The Gifford Lectures Delivered before the University of Aberdeen in the Years 1896–1898*. 4th ed. London: A. & C. Black, 1915 [1899].

Ward, Maisie. *The Wilfrid Wards and the Transition: I. The Nineteenth Century*. London: Sheed & Ward, 1934.

Ward, Wilfrid. *Problems and Persons*. London: Longmans, Green & Co., 1903.

Webb, Beatrice. *My Apprenticeship*. London, New York, and Toronto: Longmans, Green & Co., 1929.

Webb, Clement C. J. *A Study of Religious Thought in England from 1850*. Oxford: Clarendon Press, 1933.

Welch, Claude. *Protestant Thought in the Nineteenth Century: Volume I, 1799–1870*. New Haven and London: Yale University Press, 1972.

Welleck, René. *Immanuel Kant in England, 1793–1838*. Princeton, N.J.: Princeton University Press, 1931.

Werblowsky, R. J. Zwi, and Geoffrey Wigoder, eds. "Agnosticism." In *The Encyclopedia of the Jewish Religion*. New York: Holt, Rinehart & Winston, 1966, 16.

Whewell, William. *A Letter to the Author of Prolegomena Logica*. Cambridge, 1852.

White, Andrew Dickson. *The Warfare of Science*. London: Henry S. King & Co., 1876.

White, Walter. *The Journals of Walter White: Assistant Secretary of the Royal Society*. London: Chapman & Hall, 1898.

White, William Hale. *Autobiography and Deliverance*. New York: Humanities Press; Leicester: Leicester University Press, 1969 [1881].

Wiener, Philip P. *Evolution and the Founders of Pragmatism*. Cambridge: Harvard University Press, 1949.

Willey, Basil. "Honest Doubt." In *Christianity Past and Present*. Cambridge: Cambridge University Press, 1952, 107–29.

Wiltshire, David. *The Social and Political Thought of Herbert Spencer*. Oxford: Oxford University Press, 1978.

Wolff, Robert Lee. *Gains and Losses: Novels of Faith and Doubt in Victorian England*. New York and London: Garland, 1977.

Woolf, Virginia. *Moments of Being*. 2nd ed. Ed. Jeanne Schulkind. London: Hogarth Press, 1985.

Yeo, Richard. "William Whewell, Natural Theology, and the Philosophy of Science in Mid Nineteenth Century Britain." *Annals of Science* 36 (1979): 493–516.

Young, Robert M. *Darwin's Metaphor: Nature's Place in Victorian Culture.* Cambridge: Cambridge University Press, 1985.

INDEX

THE ORIGINS OF AGNOSTICISM

Designed by Ann Walston.

Composed by Action Composition Co., Inc., in Trump.

Printed by BookCrafters, Inc., on 50-lb. Booktext Natural, and bound in Holliston Roxite A with Weyerhaeuser Gainsborough end sheets.